# Springer Series in Biomaterials Science and Engineering

Volume 21

**Series Editor**

Min Wang, Department of Mechanical Engineering, The University of Hong Kong, Pokfulam Road, China

The Springer Series in Biomaterials Science and Engineering addresses the manufacture, structure and properties, and applications of materials that are in contact with biological systems, temporarily or permanently. It deals with many aspects of modern biomaterials, from basic science to clinical applications, as well as host responses. It covers the whole spectrum of biomaterials–polymers, metals, glasses and ceramics, and composites/hybrids–and includes both biological materials (collagen, polysaccharides, biological apatites, etc.) and synthetic materials. The materials can be in different forms: single crystals, polycrystalline materials, particles, fibers/wires, coatings, non-porous materials, porous scaffolds, etc. New and developing areas of biomaterials, such as nano-biomaterials and diagnostic and therapeutic nanodevices, are also focuses in this series. Advanced analytical techniques that are applicable in R&D and theoretical methods and analyses for biomaterials are also important topics. Frontiers in nanomedicine, regenerative medicine and other rapidly advancing areas calling for great explorations are highly relevant. The Springer Series in Biomaterials Science and Engineering aims to provide critical reviews of important subjects in the field, publish new discoveries and significant progresses that have been made in both biomaterials development and the advancement of principles, theories and designs, and report cutting-edge research and relevant technologies. The individual volumes in the series are thematic. The goal of each volume is to give readers a comprehensive overview of an area where new knowledge has been gained and insights made. Significant topics in the area are dealt with in good depth and future directions are predicted on the basis of current developments. As a collection, the series provides authoritative works to a wide audience in academia, the research community, and industry.

This book series is indexed by the EI Compendex and Scopus databases.

If you are interested in publishing your book in the series, please contact Dr. Mengchu Huang (Email: mengchu.huang@springer.com).

Ali Haider · Iram Shahzadi · Sehrish Kiran · Kashaf Shakoor

# Chitosan Nanoparticles for Brucellosis Treatment

Springer

Ali Haider
Department of Clinical Sciences
Muhammad Nawaz Shareef University
of Agriculture
Multan, Pakistan

Sehrish Kiran
Department of Chemistry
Lahore College for Women University
(LCWU)
Lahore, Pakistan

Iram Shahzadi
School of Pharmacy
University of Management and Technology
Lahore, Pakistan

Kashaf Shakoor
Department of Zoology, Wildlife
and Fisheries
Muhammad Nawaz Shareef University
of Agriculture
Multan, Pakistan

ISSN 2195-0644 ISSN 2195-0652 (electronic)
Springer Series in Biomaterials Science and Engineering
ISBN 978-3-032-16291-5 ISBN 978-3-032-16292-2 (eBook)
https://doi.org/10.1007/978-3-032-16292-2



This Springer imprint is published by the registered company Springer Nature Switzerland AG
The registered company address is: Gewerbestrasse 11, 6330 Cham, Switzerland

***Dedicated to***

***My Parents***

*For their support throughout my education as well as for teaching me to be independent and responsible*

***My Brothers and Sisters***

*For their love and support*

# Preface

Bovine brucellosis is a widespread zoonotic disease with significant impact on livestock production, economy, and public health, despite successful eradication as well as control programs for domestic animals established in various countries around the world. Brucellosis is one of “seven neglected endemic zoonoses” as reported by World Health Organization (WHO). *B. abortus*, *B. suis*, and *B. melitensis* are the main causative agents of bovine brucellosis from genus *Brucella* and are further subdivided into biovars. It is a significant disease of livestock instigating reproductive wastage by infertility, loss of calves, delayed heat and also contributes to economic losses owing to reduced milk as well as meat production and global trade bans. Currently, for brucellosis, no human vaccine exists, and brucellosis occurrence in the region is directly related to animal brucellosis status. Therefore, it is necessary to implement interventions that can control infection in animal reservoirs and prevent transmission of disease from animals to humans, offering more effective as well as economically viable approaches in management of disease.

Nanotechnology integration with chitosan-based systems presents a promising platform to combat such infectious diseases. Chitosan nanoparticles (CSNPs), owing to their capability to improve drug stability, controlled release, and enhance cellular uptake, offer a promising avenue for brucellosis treatment. This book offers in-depth insights, advanced applications, and current research perspectives on CSNPs for students, researchers, and professionals. The book begins with chitosan introduction, describing its origin, molecular along with chemical structure, and significant physicochemical properties. The book also provides synthesis protocol and characterization techniques of CSNPs. Bovine brucellosis is thoroughly discussed, including its impact on livestock and public health. Finally, applications of CSNPs are explored for treatment of brucellosis.


Multan, Pakistan — Ali Haider
Lahore, Pakistan — Iram Shahzadi
Lahore, Pakistan — Sehrish Kiran
Multan, Pakistan — Kashaf Shakoor

**Acknowledgments** All praises to "ALLAH," the Almighty, most Gracious, the most Merciful, and the Sustainer of the worlds, who sent us "MUHAMMAD (PBUH)" as a blessing for whole universe and the best teacher with the ultimate source of wisdom "HOLY QURAN." First and foremost, we want to express our gratitude to ALLAH, the Almighty, for bestowing upon us the health to publish our book.

We acknowledge the unwavering support of our families and colleagues whose encouragement provided the foundation for this achievement. This book is a testament to their belief in our vision. Last two authors contributed significantly in data curation, formatting, reference management, copy rights and software.

**Competing Interests** The authors have no competing interests to declare that are relevant to the content of this manuscript.

# Contents

# About the Authors

**Dr. Ali Haider** obtained his Ph.D. degree in Clinical Medicine from the Department of Veterinary Medicine (UVAS), Lahore in 2021. He served as Dairy Farm Manager at reputed commercial dairies of Punjab, (Pakistan). Later on (January, 2022–to date), Ali joined the Faculty of Veterinary and Animal Sciences, Muhammad Nawaz Shareef University of Agriculture Multan (MNS-UAM) as an Assistant Professor of Clinical Medicine. He has a robust research portfolio with over 150 international publications including patents and authored books. His research interest involves the Veterinary Clinical Medicine, Green synthesis and characterization of inorganic semiconductor nanomaterials, Polymeric nanostructures, 2D materials for their catalytic, antiparasitic and antibacterial evaluation against multiple drug resistant (MDR) pathogens.

**Dr. Iram Shahzadi** obtained her M.Phil. in Pharmaceutical Chemistry from Punjab University College of Pharmacy (PUCP), University of the Punjab, Lahore, Pakistan, in 2019 and awarded with University Gold Medal. She completed her Ph.D. from PUCP in 2024. In November 2023, Iram joined the School of Pharmacy, University of Management and Technology (UMT), Lahore, as Assistant Professor of Pharmaceutical Chemistry. Iram was recognized among the top 2% of scientists worldwide in 2024 by Stanford University and Elsevier. Her research interests encompass green synthesis, polymeric nanocomposite synthesis and characterization for water treatment, antioxidant, and anti-inflammatory activities. She also focuses on drug loading and release mechanisms for anti-cancer drugs, exploring their cytotoxicity potential against various cell lines and computational studies as in silico modeling and density functional theory (DFT) studies.

**Sehrish Kiran** obtained her B.S. (Chemistry) degree from Lahore College for Women University (LCWU), Lahore (Punjab, Pakistan) in 2020. Sehrish completed her master's degree (M.S. in Chemistry) from LCWU, Pakistan in 2022 with research work focused on removal of heavy metals from wastewater using agricultural waste. Her research interest includes synthesis and characterization of polymeric and green nanomaterials with targeted drug delivery and therapeutics, catalytic degradation and bactericidal evaluations.

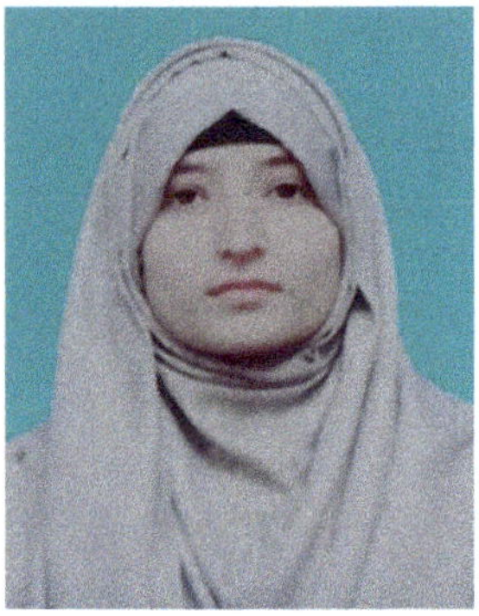

**Kashaf Shakoor** obtained her B.S. degree in Zoology from Government College University, Faisalabad (GCUF), Punjab, Pakistan in 2023. She completed M.S. Zoology from Muhammad Nawaz Shareef University of Agriculture, Multan (MNS-UAM), Pakistan (2023–25). Her research interest focuses on synthesis and characterization of polymeric nanostructures with their catalytic and bactericidal evaluations to overcome antibacterial resistance in livestock and fish industry.

# Abbreviations

| | |
|---|---|
| AFM | Atomic Force Microscopy |
| AI | Artificial intelligence |
| APCs | Antigen presenting cells |
| *B. abortus* | *Brucella abortus* |
| *B. melitensis* | *Brucella melitensis* |
| BAPA | Buffered Acidified Plate Antigen Test |
| BPAT | Buffered plate agglutination test |
| BSA | Bovine serum albumin |
| $C_8H_{13}O_5N$ | Chitin |
| CA | Calcium acetate |
| CaP | Calcium phosphate |
| CCD | Charged coupled device |
| CDC | Center for Disease Control |
| CDs | Carbon dots |
| CFT | Complement Fixation Test |
| $CH_3CONH_2$ | Acetamide |
| CMs | Chitosan microspheres |
| COS | Chitosan oligosaccharide |
| CS | Chitosan |
| CSNFs | CS nanofibers |
| CSNPs | Chitosan nanoparticles |
| CS-VLD NPs | CS-loaded Vildagliptin NPs |
| CTAB | Cetyltrimethylammonium bromide |
| CβG | Cyclic β-1,2-glucans |
| DA | Degree of acetylation |
| DCs | Dendritic cells |
| DD | Degree of deacetylation |
| $D_h$ | Hydrodynamic diameters |
| DI | Deionized |
| DL | Deep learning |
| DLS | Dynamic Light Scattering |

| | |
|---|---|
| DSe | Diagnostic sensitivity |
| DSp | Diagnostic specificity |
| EDAC | Carbodiimide hydrochloride |
| EDC | 1-ethyl-3-(3-dimethylaminopropyl) carbodiimide hydrochloride |
| ELISA | Enzyme-linked immunosorbent assay |
| Eud | Eudragit |
| FNP | Flash nanoprecipitation |
| FTIR | Fourier-transform infrared spectrophotometry |
| GA | Glutaraldehyde |
| GA | Gum arabic |
| GCA | Glyceraldehyde |
| GCM | Group contribution method |
| Gd-DTPA | Gadopentetic acid |
| GlcNAc | N-acetyl-2-amino-2-deoxy-D-glucose |
| GO | Graphene oxide |
| $HAuCl_4$ | Hydrogen tetrachloroaurate |
| i.p. | Intraperitoneally |
| I-Carrageenan | Iota-Carrageenan |
| IgA | Immunoglobulin A |
| IPN | Interpenetrating polymer network |
| $K_a$ | Dissolution constant |
| $La_2O_3$ | Lanthanum oxide |
| LFIA | Lateral flow immunochromatographic assays |
| LPS | Lipopolysaccharide |
| MCN | Mannosylated Chitosan Nanoparticles |
| Mdh | Malate dehydrogenase |
| MEs | Microemulsion systems |
| MHS | Mark Houwink Sakurada |
| ML | Machine learning |
| $MnO_2$ | Manganese dioxide |
| M-RT-PCR | Multiplex real-time PCR |
| Msat | Micro-standard agglutination test |
| MW | Molecular weight |
| MWD | Molecular weight distribution |
| $MW_v$ | Viscosity-average molecular weight |
| $MW_w$ | Weight average molecular mass |
| NALT | Nasal Associated Lymphoid Tissue |
| NaOH | Sodium hydroxide |
| NGS | Next-generation sequencing |
| NHS | N-hydroxysuccinimide |
| NLCs | Nanostructured lipid carriers |
| NO | Nitric oxide |
| NPs | Nanoparticles |
| NTA | Nanoparticle Tracking Analysis |
| PA | Phosphoric acid |

| | |
|---|---|
| PBS | Phosphate buffered saline |
| PCL | Polycaprolactone |
| PCR | Polymerase chain reaction |
| PCR-EIA | PCR-enzyme immunoassay |
| PCS | Photon Correlation Spectroscopy |
| PDI | Polydispersity index |
| PECs | Polyelectrolyte complexes |
| PEO | Polyethylene oxide |
| PET | Polyethylene terpthalate |
| PGF2α | Prostaglandin F 2-alpha |
| PLA | Polylactic acid |
| PNPs | Polymeric nanoparticles |
| PPE | Personal protective equipment |
| PVA | Polyvinyl alcohol |
| PVP | Polyvinylpyrrolidone |
| QbD | Quality by Design |
| RBT | Rose Bengal test |
| Riv.T. | Rivanol Test |
| RMS | Root mean square |
| ROS | Reactive oxygen species |
| SAT | Standard agglutination test |
| SEM | Scanning Electron Microscopy |
| Semi-IPN | Semi-interpenetrating networks |
| SLNs | Solid lipid nanoparticles |
| S-LPS | Smooth lipopolysaccharide |
| SnSe | Stannous selenide |
| STPP | Sodium tripolyphosphate |
| T4SS | Type IV secretion system |
| TEGDMA | Triethylene glycol dimethacrylate |
| TEM | Transmission Electron Microscopy |
| TFA | Trifluoroacetic acid |
| THF | Tetrahydrofuran |
| TIR | Toll/interleukin-1 receptor |
| TLR2 | Toll like receptor 2 |
| TMC | N-trimethyl chitosan |
| W/O MEs | Water in oil microemulsion systems |
| w/v | Weight/volume |
| WHO | World Health Organization |
| ζ-potential | Zeta potential |

# List of Figures

# List of Tables

# Chapter 1
# Chitosan Introduction

**Abstract** Chitosan (CS), a naturally occurring polysaccharide obtained through partial deacetylation of chitin, has emerged as a multifunctional biomaterial owing to its distinctive structural, functional, and chemical properties. The reactive amino and hydroxyl groups present in CS enable different chemical modifications as derivatization, which facilitate precise alteration of biological and physicochemical properties, ideally adopting any application field. Hydrophilic or hydrophobic derivatization of CS enhances hydrogel formation potential, anticoagulant, antibacterial and hemostatic activities. The integration of CS chains with graphene oxide (GO) nanosheets is an effective approach to create hydrogels that exhibit significant potential for applications in biomaterials and as creative soft materials within environmental sciences. Fundamental technological attributes, including solubility behavior and viscosity, determine the influence of molecular characteristics and processing conditions on the performance of CS. Moreover, its biological activities as antibacterial, antioxidant, and anti-inflammatory properties enhance its multifunctional potential. The antibacterial actions of chitosan nanoparticles (CSNPs) are highlighted considering their potential application in combating brucellosis. Collectively, these aspects highlight the significant potential of CS and its derivatives for applications in advanced technology and biomedical fields.

**Keywords** Antibacterial · Chitosan · Graphene oxide · Hydrogel · Nanoparticles

## 1.1 Introduction

Polysaccharide polymers derived from natural origins emerging in pharmaceutical technology attributed to their less toxicity, biodegradability, and biocompatibility. Chitosan (CS) is a naturally occurring polysaccharide after cellulose, is a biodegradable and biocompatible mucoadhesive polymer, widely employed in synthesis of nano- as well as microparticles [1]. CS is an altered biopolymer obtained via partial chitin deacetylation. Chitin is a widely occurring biopolymer present in exoskeleton of shrimp shells, crab, crustacea, insect's cuticles, algae, and in cell wall of fungi.

 
A. Haider et al., *Chitosan Nanoparticles for Brucellosis Treatment*, Springer Series in Biomaterials Science and Engineering 21, https://doi.org/10.1007/978-3-032-16292-2_1

CS is comprised of alternating units of glucosamine (1 → 4) connected to N-acetyl glucosamine. CS is a hard, white, nitrogenous, and inelastic polysaccharide.

Chitin was first discovered by Henri Braconnot in 1811 while researching mushrooms. Professor C. Rouget in 1859 discovered that chitin treatment with alkali produced a substance that can dissolve in acids unlike chitin. The deacetylated chitin is termed as CS by Hoppe Seiler [2, 3]. CS present in some fungi is less common in nature. In the past, CS samples were primarily produced from crustacean sources by chitin deacetylation. Lately, CS from fungi is becoming popular in the market due to vegan demands. These samples are controlled properly with regard to low viscosity and possess extraordinary degrees of deacetylation. Production of CS from cuticles of insects is also garnering attention driven by increasing focus on production of protein from these sources [3, 4].

The annual production of biosynthesized chitin is approximately 10 billion tons. The preparation process involves the steps of demineralization, deproteinization, decoloration, and deacetylation, transforming crustacean shells into CS. Chitin consists of 5–8% weight/volume (w/v) nitrogen primarily in the form of significant aliphatic amino groups as present in CS depending on deacetylation extent. CS structure has greatest metal-coordinating capacity in adsorption among natural polymers attributed to presence of reactive functional groups like primary along with secondary hydroxyl and amino/acetamido group [4].

Recent advancements in fermentation technology facilitate the regulated production of CS through the cultivation of microorganisms, such as fungi, which incorporate CS within their cell walls. *Gongronella butleri* can give the highest chitosan yield among various *Zygomycetes* fungi including *Mucor*, *Absidia*, *Rhizopus*, and *Gongronella.* Solid-substrate fermentation is an excellent method to obtain CS through fungus *G. butleri* [5]. Chitin is associated with melanin in insects, a feature that is not present in crustaceans. Moreover, insects may possess significant fat content. Consequently, additional purification procedures are necessary in the process of chitin extraction from insects. Despite these challenges, there has been a notable increase in recent studies focused on the extraction of chitin and its derivatives from insects. This pertains to the cultivation of insects on a grand scale and the management of exotic species [6].

Recently, nanotechnology has gained attention in various fields due to numerous up-and-coming novel usages [7]. Nanotechnology is applied in developing innovative drug delivery methods by drug entrapment in nanoparticulate systems [1]. Particles that are sub-microscopic and colloidal in nature, measuring less than 100 nm in dimension, are referred to as nanoparticles (NPs). NPs have emerged as highly promising vectors, demonstrating a broad spectrum of applications in the realms of targeted gene and drug delivery. NPs offer optimal targeting capabilities and enhanced tissue penetration because of their diminutive scale. Engineered NPs derived from a range of synthetic and natural polymers, such as CS and polyethylenimine, have been extensively investigated for their potential in delivering anticancer drugs and nucleic acids like DNA and siRNA. The primary classifications of NPs are nanospheres and nanocapsules. The distinction lies in the structure of nanospheres, which are spherical particles that can either encapsulate drugs within their interior or adsorb them onto their exterior. In contrast, nanocapsules consist of a solid polymeric shell that encases a liquid core, with drugs being either adsorbed on the outer surface or contained within the core itself. NPs can be observed in different shapes such as cylinders, spheroids, cones, nanorods, nanotubes, etc. [8] NPs possess exceptional properties as compared to their bulk counterparts owing to decline in dimension to atomic scale. The properties of materials undergo significant alterations at the nanoscale, as bulk materials maintain consistent characteristics regardless of their dimensions. However, as the size diminishes, the proportion of surface atoms relative to the bulk material escalates. This leads to unexpected properties of NPs [3].

Chitosan nanoparticles (CSNPs), as advantageous polymeric and bio-based nanomaterials, have garnered significant interest in recent decades [9]. CSNPs were initially introduced in 1994, when the intravenous administration of 5-fluorouracil, an anticancer agent, was suggested to be facilitated by CSNPs obtained through cross-linking and emulsification techniques. Moreover, various scholars have developed innovative formulations of CSNPs that incorporate secondary matrix-forming substances. A diverse array of methodologies has been introduced for the formation of CSNPs, primarily encompassing emulsification, various forms of coacervation, or even subtle modifications thereof. The methods in question encompass emulsion droplet coalescence, ionic gelation, the reverse micellar approach, emulsion solvent diffusion, desolvation, and polyelectrolyte complexation. All these techniques are categorized as bottom-up fabrication processes, wherein molecules aggregate in solution to create NPs [10]. CS can be transformed into various forms such as gels, solutions, sponges, blends, membranes, tablets, paste, microgranules, or microspheres for a multitude of applications [11].

CS is a non-toxic and biodegradable polymer, shows low level of immune reactions, possesses mucoadhesive along with absorption-enhancing attributes. Owing to beneficial properties, CS is widely employed in drug delivery systems with increased sensitivity and specificity, improved biodistribution, and reduced pharmacological toxicity. CSNPs have excellent anticancer properties confirmed by several researchers. CS is characterized by greater biocompatibility along with poor toxicity because of its natural structural affinity for glycosaminoglycan. It is easily biodegradable into non-toxic metabolites which are completely absorbed by the human body. CS impedes the function of efflux pumps, thereby obstructing the transport of certain proteins located on the membranes of enterocytes or intestinal epithelial cells that release xenobiotics, primarily drugs. Consequently, these transporters play a crucial role in the mechanisms underlying drug resistance. CSNPs serve as exceptional drug carriers due to several intrinsic advantageous properties of CS, including straightforward preparation, bioactivity, non-toxic characteristics, biocompatibility, biodegradability, and a cationic nature. The favorable mucoadhesive properties of CSNPs enhance bioavailability and the dissolution rate of hydrophobic drugs. Furthermore, CS contributes to the stability, bioavailability, and controlled release of unstable drugs, as the molecules exceed the nanoscale range, thereby effectively surmounting barriers and enhancing permeability [12]. CS is commonly employed to prepare hydrogels for drug delivery owing to its important characteristics as bio-adhesion, having a polycationic surface that facilitates formation of ionic along with hydrogenic bonds, and its capability of not producing any toxins or triggering immune response while contacting with living tissue or body fluids [13]. CS is widely employed in biomedical industries, genetic engineering, agriculture, controlling environmental pollution, food industry, paper manufacturing, photography, water treatment, and so on [3]. The removal of organic contaminants such as PTCs, dyes, and pharmaceuticals from wastewater can be effectively achieved through the utilization of nanomaterials composed of CS [14].

## 1.2 Chitosan Chemistry

Chitin ($C_8H_{13}O_5N$) is derived from the Greek word "chiton" meaning "coat of mail." This polysaccharide is composed of –(1 → 4)–N acetyl-D-glucosamine monomers present in nature. Chitosan is a chitin variant that was reported about 40 years after chitin identification. Chitin and cellulose are often mistaken because of their very similar chemical structures, even though chitin possesses acetamide groups ($CH_3CONH_2$) at the carbonyl group rather than the hydroxyl group. The crystalline structure of chitin consists of poly (1, 4)-linked N-acetyl-2-amino-2-deoxy-D-glucose (GlcNAc), interspersed with a limited number of 2-amino-2-deoxy-D-glucose residues. The material exhibits a white, inelastic, nitrogenous composition, characterized by its toughness. The three crystalline allomorphic forms of chitin include α-chitin, β-chitin, and γ-chitin, which can be observed in nature. α-chitin comprises anti-parallel chains that contribute to the rigidity of the polymer. The majority of shrimp exoskeletons and crab structures are composed of α-chitin. Nonetheless, β-chitin consists of parallel chains that create monocyclic crystals, characterized by both intermolecular and intramolecular interactions through hydrogen bonds. For instance, squid pens, diatom spines, and pogonophoran tubes are composed of β-chitin. γ-chitin is a composition of both parallel and anti-parallel chains, exhibiting properties akin to those of α-chitin and β-chitin, and can be found in insect cocoons, fungi, and yeasts. CS, a linear polysaccharide, is derived from acetylated and deacetylated d-glucosamine units connected by (1, 4) glycosidic bonds (Fig. 1.1). Chitin deacetylation results in the formation of $–NH_2$ groups and acetate ions, which are produced through the hydrolysis of the acetamide group [15]. When repeating units of chitin lack the acetyl functional structure, specifically as β-1,4-D-glucosamine following deacetylation, the resulting polymer is referred to as CS [16]. CS exhibits a distinctive cationic characteristic, as most polysaccharides tend to be neutral or negatively charged in acidic conditions. This property allows CS to interact with other negatively charged natural or synthetic polymers, resulting in the formation of multilayer structures or electrostatic complexes [17].

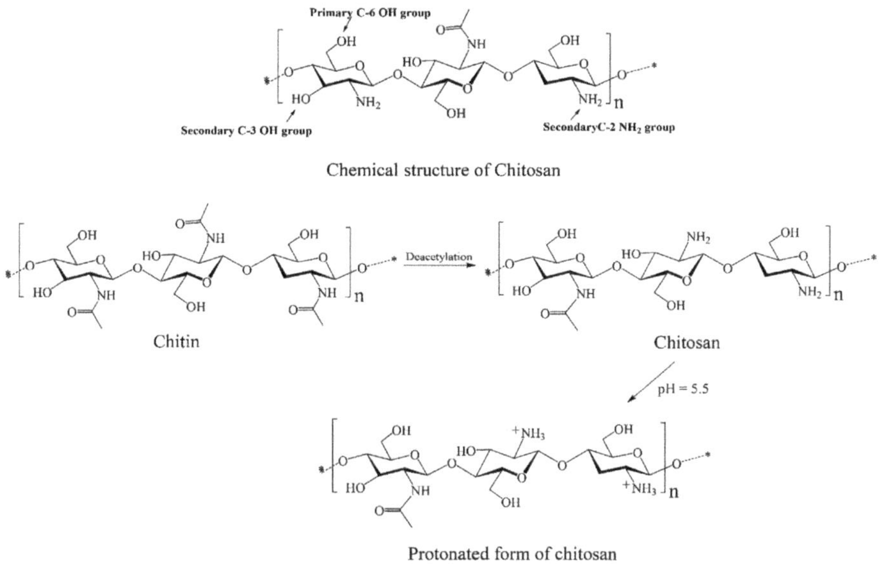


**Fig. 1.1** Structure of chitin, chitosan, and protonated form of chitosan. Reproduced with permission from Ref. [18] Copyright 2017, American Chemical Society

## *1.2.1 Chemical Modification of Chitosan*

The biological and physicochemical properties of CS can be tuned through chemical modification ideally adopting any application field. Two hydroxyl groups at C-3 (secondary) and C-6 (primary) on Glc*N*Ac and Glc*N* units and two amino groups distinctive to glucosamine units of CS can be chemically modified (Fig. 1.2). Amines are more reactive than hydroxyls and both amino and functional hydroxyl groups might react with electrophilic reagents as nucleophiles. Robust chemistry that allows for strict control over selectivity is essential for CS derivatization, as regioselective and chemo-selective issues may arise during CS modification. A lack of selectivity can lead to heterogeneous mixtures of derivatives, which limit applications and reproducibility [19].

Moreover, aqueous acids are required as solvents for CS functionalization, which works effectively in homogenous conditions. Physicochemical properties of CS derivatives can be tuned by introducing various functional groups owing to the unique chemical nature of the acyl or alkyl group of derivatizing agents (Fig. 1.3). In particular, CS derivatives exhibiting variable properties based on substitution degree can be synthesized by incorporating hydrophilic groups such as carboxylic acids, sulfonates, sulfates, or phosphate groups in their anionic form at appropriate pH levels. As a result, these derivatives exhibit solubility in water at neutral pH and demonstrate enhanced coordination capabilities with metal cations, hydrogel formation potential, adjustable viscosity, and a range of biological activities, including anticoagulant and antibacterial properties. Amphiphilic derivatives that enhance surface-active proteins

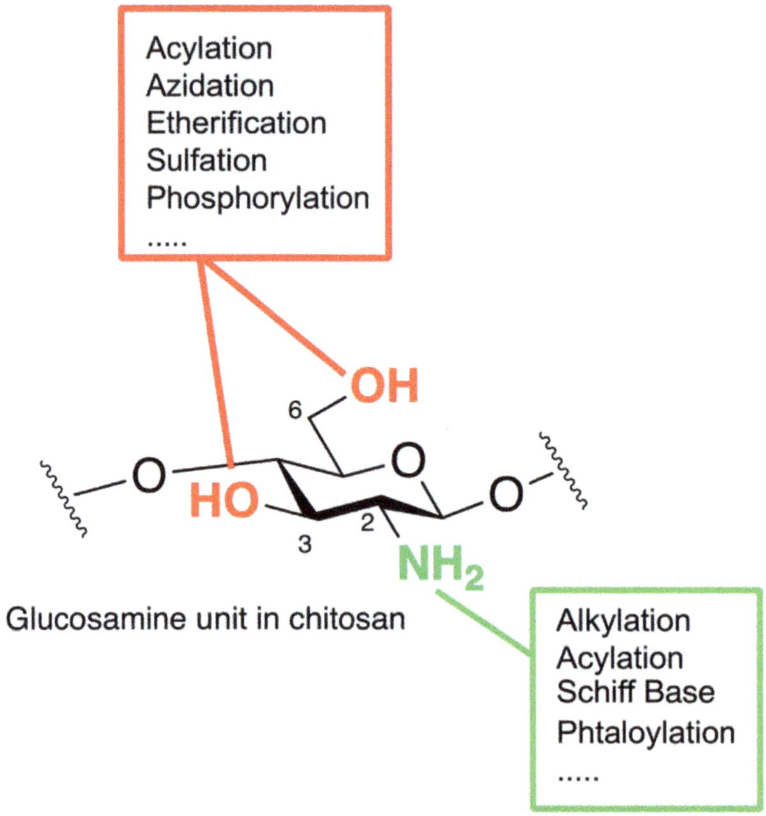


**Fig. 1.2** Sites of derivatization of Glc*N* units in CS. The modification of Glc*N*Ac unit is limited to 3- and 6-OH. Reproduced with permission from Ref. [19]

and influence aggregation behavior can be synthesized by incorporating hydrophobic groups into CS, achieved through alkylation with hydrocarbon chains or acylation with fatty acids, leading to the formation of hydrogels, nanoparticles, and micelles. These derivatives have significant applications in the biomedical industry, particularly in drug delivery systems and the absorption of lipophilic water pollutants such as oils. Moreover, hydrophobic derivatization enhances both antibacterial and hemostatic activities [19].

### *1.2.2 Chitosan Chains Incorporation with Graphene Oxide (GO) Nanosheets*

Graphene is a single-atom-thick nanosheet that belongs to the family of carbon nanomaterials, organized in a honeycomb two-dimensional lattice. It is utilized in a wide

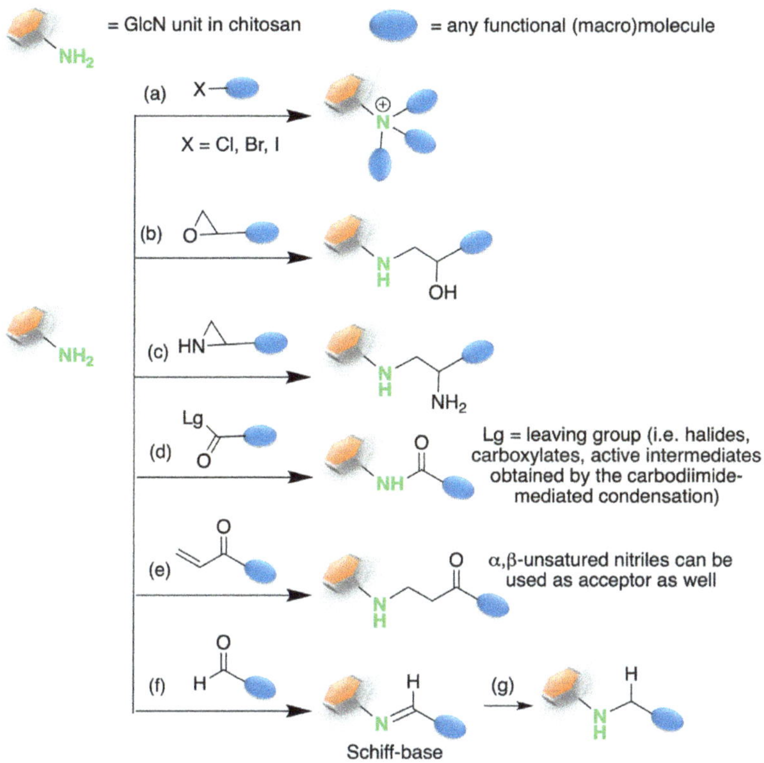


**Fig. 1.3** Chemistry of GlcN unit modifications at the 2-amino group; **a**, **b**, **c** alkylation; **d** acylation; **e** Michael addition; **f**, **g** reductive amination. Reproduced with permission from Ref. [19]

range of applications due to its unique electronic and mechanical properties. Recently, supramolecular hydrogels composed of CS and GO have been synthesized through non-covalent interactions, with GO nanosheets serving as two-dimensional cross-linkers. Hydrogel can be synthesized at room temperature with high concentrations of GO, whereas supramolecular hydrogel formation occurs at 95 °C under low GO concentrations. These novel hydrogels exhibit significant potential for applications in biomaterials and as creative soft materials within environmental sciences [20].

## 1.3 Technological Chitosan Properties

### *1.3.1 Solubility*

CS is generally prepared via $\alpha$-chitin deacetylation by 40–50% aqueous alkali solution at 100–160 °C for a few hours. Obtained CS possesses a degree of deacetylation (DD) up to 0.95. Alkaline treatment is repeated to attain complete deacetylation. CS is an N-deacetylated chitin derivative with a typical degree of acetylation (DA) less than 0.35. Physical properties of CS depend upon parameters such as molecular weight (MW) from 10,000 to 1 million Dalton, DD between 50 and 95% product purity, and sequence of amino along with acetamido group and molecular weight (MW) approximately from 10,000 to one million Dalton [21].

Due to the semi-crystalline chitin structure with vast hydrogen bonding, cohesive energy density and solubility parameter are very high; and consequently, it is insoluble in all usual solvents. Solubility parameter of CS and chitin can be determined through group contribution methods (GCM). CS is freely soluble in diluted acidic solutions under pH 6.0, whereas chitin is insoluble in various organic solvents. This phenomenon can be attributed to the robust basicity of CS, which is characterized by the presence of amino groups exhibiting a $pK_a$ value of 6.3. The existence of amino groups indicates that pH significantly influences the charged state and characteristics of CS. Amines undergo protonation, resulting in a positive charge at low pH, thereby rendering CS a water-soluble cationic polyelectrolyte. In contrast, as the pH rises over 6, the amines deprotonate, the polymer loses its charge, and it becomes insoluble.

Dissolution constant ($K_a$) of amine group derived from equilibrium as:

$$NH_2 + H_2O \leftrightarrow -NH_3^+ + OH^-$$
$$K_a = [-NH_2][H_3O_-]/\left[NH_3^+\right] \text{ and } pK_a = -\log K_a$$

For polyelectrolytes, the dissociation constant depends on the dissociation degree at which it can be determined [22]. DD is defined as the mole fraction of repeating units with free $-NH_2$ groups in polysaccharides, while DA is defined as the percentage of *N*-acetylated repeating units. Thus, the relationship between DD and DA is defined as DA + DD = 100% [16]. Commercial CS generally has molar weight in the range of $10^4$–$10^6$ g mol$^{-1}$ [23]. DA is a significant parameter because CS with lower DA displays robust biological effects as well as water solubility. For instance, previous studies discovered that CS with DA less than 50% can be water soluble because lesser DA recommends higher $-NH_2$ groups concentration in the molecule, and protonation of amino groups is essential for water solubility and biological effects.

Correlation between CS molecular weight (MW) and water solubility is the same as the connection between biological activities and MW. Solubility increases when MW decreases. CS with $MW_v$ less than 30 kDa is water soluble without using acid in free amine form. However, CS with $MW_w$ from 22 to 30 kDa only shows limited solubility, whereas CS with $MW_w$ below 9 kDa displays remarkably best solubility in water. When $MW_v$ is above 30 kDa, amino group protonation via applying acid is essential to dissolve CS in water. Acetic acid is commonly employed for this purpose; however, other acids found within the human body, such as HCl, pyruvic acid, citric acid, and lactic acid, are also capable of solubilizing CS in water, with the exception of phosphoric acid. The outcome of this reaction is the formation of CS hydrochloride as the final product. It is important to observe that CS with a molecular weight exceeding 30 kDa is soluble exclusively in aqueous solutions where the pH is below 6.0 [16, 24, 25].

### *1.3.2 Viscosity*

CS produce viscous solutions in acidified aqueous solvents. A minor increase in the concentration of CS significantly elevates viscosity. The high surface charge density of CS is due to its polyelectrolytic characteristics, which facilitate the formation of extensive intermolecular and intramolecular hydrogen bonds through amine and hydroxyl groups, leading to the development of viscous solutions. Furthermore, the viscosity of CS is influenced by the degree of deacetylation, which is dependent on surface charge. The high degree of deacetylation increases the number of amine groups, which enhances hydrogen bonding and consequently raises viscosity. Moreover, a high degree of deacetylation increases the charge density on CS molecules, resulting in significant intramolecular repulsive forces. This phenomenon enhances rigidity and promotes molecular expansion of individual polymer chains, consequently increasing viscosity (Fig. 1.4) [26].

The effectiveness of CS is contingent upon MW and molecular weight distribution. The rheological and mechanical properties can be determined if the MW and polymer conformation are well understood. The average MW distribution (*Mn, Mw, or Mz*) exhibits chronophagic characteristics. The determination of viscosity-average ($MW_v$) is a relatively rapid method. The intrinsic viscosity of a polymer solution is related to the polymer MW as described by the Mark–Houwink–Sakurada (MHS) equation:

$$[\eta] = KM_v^a$$

where $\eta$ is intrinsic viscosity, $M_v$ is viscosity-average MW, $K$ and a are constants for given solute–solvent system [27].

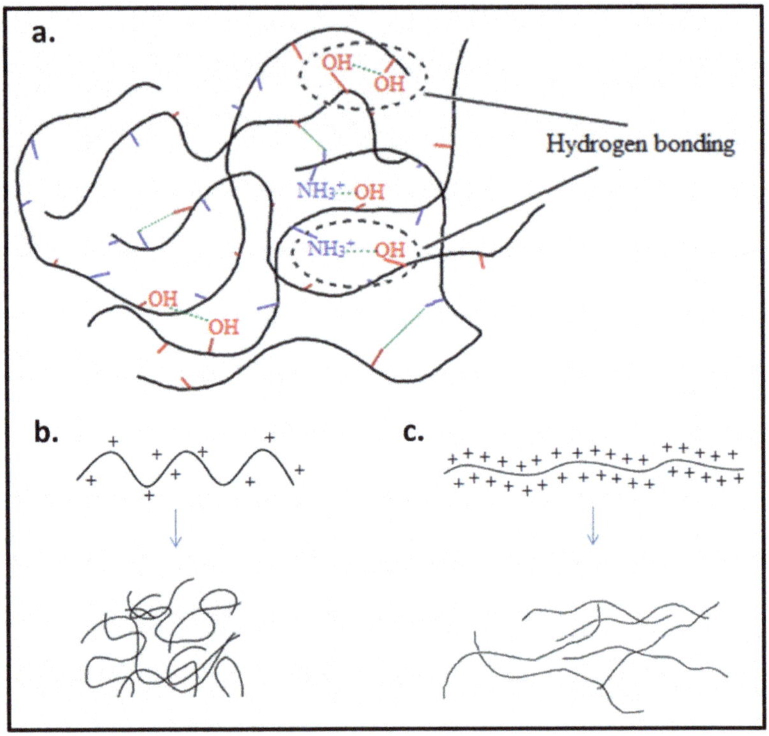


**Fig. 1.4** **a** Extensive inter and intramolecular hydrogen bonding among amine and the hydroxyl groups present in chitosan molecules (leads to a significant increase in viscosity with a slight increase in the concentration of chitosan); **b** Chitosan molecule with low surface charge density (e.g., at high pH) exhibiting less intermolecular repulsion resulting in contracted molecules and low viscosity; **c** Chitosan molecule with high charge density (e.g., at low pH) showing strong inter and intramolecular repulsion, high stiffness, considerable expansion, and high viscosity. Reproduced with permission from Ref. [26]. Copyright 2021, Elsevier

## 1.4 Biological Properties

### *1.4.1 Antibacterial Activity*

Brucellosis has been recognized as a zoonotic disease with great economic and animal welfare ramifications globally known since prehistoric times. Seven *Brucella* species impacting terrestrial animals are *Brucella abortus* (*B. abortus*), *Brucella melitensis* (*B. melitensis*), *Brucella ovis* (*B. ovis*), *Brucella suis* (*B. suis*), *Brucella neotomae* (*B. neotomae*), *Brucella canis* (*B. canis*), and *Brucella microti* (*B. microti*), while *Brucella pinnipedialis* (*B. pinnipedialis*) and *Brucella ceti* (*B. ceti*) distress marine mammals. There are 7 biovars of *B. abortus*, 5 for *B. suis*, and 3 for *B. melitensis*. Brucellosis treatment through antibiotics in domestic animals is repeatedly ineffective due to intracellular survival and adaptability of *Brucella* in macrophages. Low efficacy of treatment and reinfection are most common in humans. The selection of drug combinations for brucellosis treatment should be conducted with caution to mitigate side effects in humans and the emergence of resistance [28].

There has been a significant rise in the utilization of both organic and inorganic nanomaterials, encompassing metal/metal oxide and polymeric NPs, in the formulation of traditional antibiotics. These innovative formulations exhibit noteworthy antibacterial properties, paving the way for new strategies in the battle against microbial infestations [29]. The immunogenicity of omp19 antigens delivered by CSNPs in *B. melitensis* and *B. abortus* has been thoroughly examined. The production of antigens in the form of NPs can elicit Th1 and Th2 responses in laboratory samples. The administration of oral antigens has the capacity to elicit Th1 and Th17 responses. The utilization of these antigens as NPs serves to elicit immune responses and confer protection against *B. melitensis* and *B. abortus* [30].

Chitosan microspheres (CMs) are efficient carriers to encapsulate proteins and might attain high protein loading efficacy and protect proteins from degradation (Fig. 1.5). At present, numerous protein-loaded CMs are effectively developed. Intranasal delivery of CMs-BLSOmp31 can induce systemic and mucosal immune responses preventing colonization and disease in sheep by *B. ovis*. The bioactivity of nerve growth-factor can be preserved in CSMs, rendering them promising candidates for the encapsulation of growth factors. The cationic polysaccharide CS exerts a temporary influence on the permeability of tight junctions. Consequently, CS may impede the rate of mucociliary clearance through its interactions with the nasal mucosa, thereby prolonging the duration for immunogen absorption. Both properties facilitate the interaction and absorption of antigens by immunocompetent Nasal Associated Lymphoid Tissue (NALT) cells, which play a crucial role in the activation of mucosal immune responses, resulting in robust and sustained local cellular and humoral immune responses [31].

Live attenuated vaccines play a vital role in the prevention of animal brucellosis. CS and its derivatives have been utilized in prior research as nanoparticulated adjuvants for the delivery of antigens in subunit vaccines. The heightened expression of mannose receptors on the surface membranes of antigen-presenting cells (APCs), such as macrophages and dendritic cells, facilitates the binding of mannose residues found on antigens through specific ligand-receptor interactions. This process enables the transfer of antigens into endocytic pathways, thereby initiating MHC presentation and subsequent T cell activation. Consequently, the mannosylation of CSNPs can trigger receptor-mediated endocytosis aimed at targeting APCs, thereby augmenting the capacity for antigen presentation and improving the efficacy of immune responses. Mice immunized with protein FliC and FliC-MCN (mannosylated CSNPs) provide remarkable protection against *B. melitensis* 16M and *B. abortus* 544 infestations. Conversely, live attenuated *B. abortus* RB51 and *B. melitensis* Rev1 vaccination groups provide most superior protection regarding *B. abortus* 544 and *B. melitensis* 16 M [32].

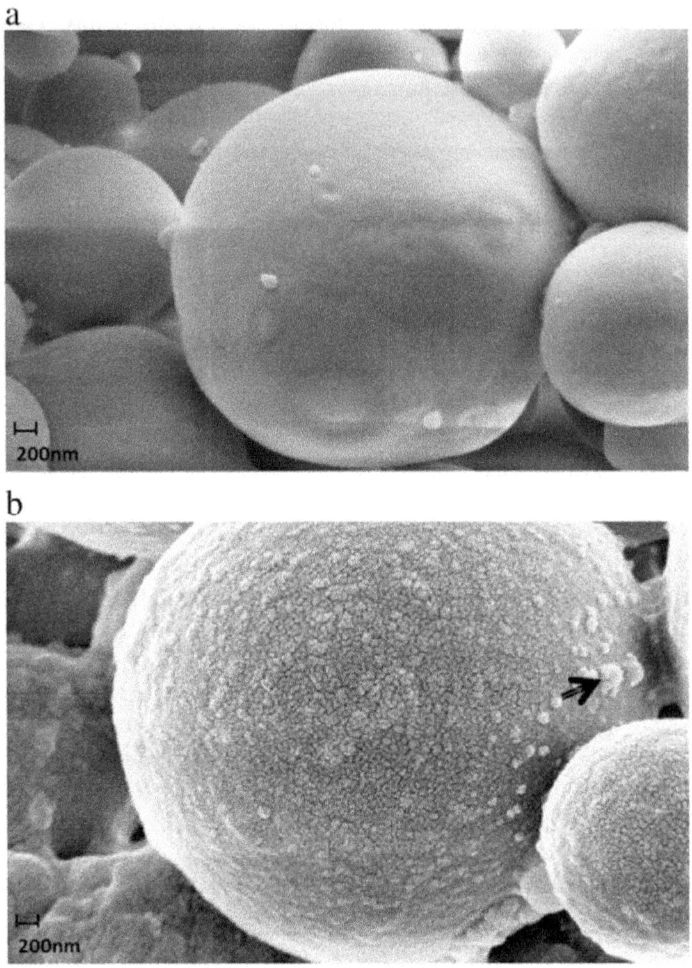


**Fig. 1.5** Scanning electron micrographs of chitosan microspheres prepared by spray-drying after dispersed in deionized water and dried at 45 °C: **a** blank chitosan microspheres and (73450X). **b** BLSOmp31-loaded chitosan microspheres (120990X). Black arrow indicates BLSOmp31 on surface of a microsphere. Reproduced with permission from Ref. [31]. Copyright 2021, Elsevier

## *1.4.2 Antioxidant Activity*

Human metabolism generates free radicals within the body, primarily in the form of reactive oxygen species (ROS) such as hydrogen peroxide ($H_2O_2$), hydroxyl radicals, and superoxide anions. Reactive oxygen species induce oxidative harm to cellular structures through their interaction with DNA, lipids, and proteins, while also instigating inflammation that may lead to a variety of additional pathologies. Antioxidant compounds effectively neutralize free radicals, which can be attributed to their beneficial effects. CS demonstrates antioxidant capabilities against DPPH radicals, yielding a value of 24.6 ± 1.73%, and against ABTS radicals, exhibiting a value of 84.1 ± 1.42% accordingly [33]. CS exhibits the ability to scavenge free radicals and chelate metal ions, thereby inhibiting hydrogen donation and contributing to its antioxidant properties. Furthermore, the diminutive dimensions and low molecular weight of NPs enhance the antioxidant capacity of CSNPs. The antimicrobial properties of CSNPs have been attributed to their capacity for antioxidant activity [34]. Antioxidants serve as protective agents that counteract the detrimental effects of excessive oxidation. Oxidative stress arises from an imbalance in the antioxidant status within humans [35].

The scavenging activity of CS can be attributed to its robust capacity for hydrogen donation. Reactive oxygen species interact with hydrogen atoms present in hydroxyl or amino groups of carbon structures, resulting in the formation of stable macromolecular radicals. CS exhibits remarkable metal bonding potential attributed to the presence of free amino groups. The beneficial impact of low MW and higher concentration on the antioxidant activity of CS is noteworthy. Conversely, elevated MW serves as a significant factor in reducing lipid oxidation. The antioxidant properties exhibited by various MWs of CS can be ascribed to their ability to chelate metal ions. High MW CS should exhibit reduced mobility compared to lower MW CS, which may enhance the potential for both inter- and intramolecular bonding between metals and high MW CS. This may account for reduced chelation capacity due to the high MW of CS. Numerous studies have examined CS independently as a potential antioxidant agent. The scavenging of hydroxyl radicals by CS inhibits the lipid peroxidation of phosphatidylcholine in conjunction with linoleate liposomes. The various DD of CS exhibit significantly enhanced antioxidant activities, ranging from 58.3 to 70.2% at a concentration of 1 mg/ml, and achieving levels between 79.9 and 85.2% at 10 mg/ml [36].

### *1.4.3 Anti-inflammatory Properties*

Inflammation represents the initial defensive response to injury or infection in the human body, initiated within tissue compartments by a specific array of inflammatory and immune cells, with the objective of restoring both structural and functional integrity after exposure to harmful stimuli. The anti-inflammatory activity of CS is primarily attributed to the structural elements of its molecule rather than its MW [37]. Inflammation serves as a fundamental cause of numerous invertebrate diseases, including heart disease, autoimmune disorders, and inflammatory bowel disease, thereby complicating the treatment of chronic conditions. The prevailing constraint in treatment lies in the inadequate selection of pharmaceuticals when distinguishing between normal and inflammatory tissues. Therefore, to address the aforementioned issue, research is being conducted to develop drug carriers that utilize the targeted regions of damaged tissues to enhance the accumulation of drugs within inflammatory cells, thereby facilitating controlled drug release and minimizing the concentration of the drug in healthy tissues. As a result, it will reduce the adverse effects of the medication and enhance the efficacy of the treatment [38].

CS is recognized for its anti-inflammatory properties and gelatine is well known for pro-inflammatory activities. According to earlier studies, CS can change activities of macrophages. Cryogel scaffolds based on CS created with different gelatin proportions were tested for immunomodulatory activities in presence of lipopolysaccharide (LPS) as stimulant. They functioned as anti-inflammatory molecules based on levels of TNFα production. When the CS ratio exceeded that of gelatin, the materials exhibited anti-inflammatory activity in powdered forms, utilizing TNFα production levels as a benchmark from macrophages activated by LPS. However, an increased gelatin ratio negated this effect [39].

Low viscosity and higher solubility of chitosan oligosaccharide (COS) at neutral pH have gained attention of most researchers to utilize CS in oligosaccharide form. Exposure to COS results in dose-dependent reduction of LPS-induced TNF-α and IL-6 secretion in incubation medium. Furthermore, corresponding reduction in TNF-α and IL-6 at mRNA showed that COS exposure downregulated expression of cytokines at transcription level. COS exposure can also decrease LPS-induced nitric oxide (NO) secretion in medium. Interestingly, external TNF-α addition to medium reversed COS-mediated decline in IL-6 and NO levels thus representing that anti-inflammatory effect of COS occurs by TNF-α pathway modulation [40].

## References

1. Ahmed TA, Aljaeid BM (2016) Preparation, characterization, and potential application of chitosan, chitosan derivatives, and chitosan metal nanoparticles in pharmaceutical drug delivery. Drug Des Dev Ther 10:483–507
2. Aranaz I et al (2021) Chitosan: an overview of its properties and applications. Polymers 13(19):3256
3. Divya K, Jisha M (2018) Chitosan nanoparticles preparation and applications. Environ Chem Lett 16:101–112
4. Liu X-Q et al (2022) Review on preparation and adsorption properties of chitosan and chitosan composites. Polym Bull 79(4):2633–2665
5. Huq T et al (2022) Sources, production and commercial applications of fungal chitosan: a review. J Bioresour Bioproducts 7(2):85–98
6. Khayrova A, Lopatin S, Varlamov V (2021) Obtaining chitin, chitosan and their melanin complexes from insects. Int J Biol Macromol 167:1319–1328
7. Chandrasekaran M, Kim KD, Chun SC (2020) Antibacterial activity of chitosan nanoparticles: a review. Processes 8(9):1173
8. Chandra Hembram K et al (2016) Advances in preparation and characterization of chitosan nanoparticles for therapeutics. Artif Cells Nanomed Biotechnol 44(1):305–314
9. Yanat M, Schroën K (2021) Preparation methods and applications of chitosan nanoparticles; with an outlook toward reinforcement of biodegradable packaging. React Funct Polym 161:104849
10. Grenha A (2012) Chitosan nanoparticles: a survey of preparation methods. J Drug Target 20(4):291–300
11. Boroumand H et al (2021) Chitosan-based nanoparticles against viral infections. Front Cell Infect Microbiol 11:643953

12. Jafernik K et al (2023) Chitosan-based nanoparticles as effective drug delivery systems—a review. Molecules 28(4):1963
13. Sachdeva B et al (2023) Chitosan nanoparticles-based cancer drug delivery: application and challenges. Mar Drugs 21(4):211
14. Kandasamy G, Manisekaran R, Arthikala M-K (2024) Chitosan nanoplatforms in agriculture for multi-potential applications-adsorption/removal, sustained release, sensing of pollutants & delivering their alternatives—a comprehensive review. Environ Res 240:117447
15. Iber BT et al (2022) A review of various sources of chitin and chitosan in nature. J Renew Mater 10(4):1097
16. Kou SG, Peters L, Mucalo M (2022) Chitosan: a review of molecular structure, bioactivities and interactions with the human body and micro-organisms. Carbohyd Polym 282:119132
17. Cheung RCF et al (2015) Chitosan: an update on potential biomedical and pharmaceutical applications. Mar Drugs 13(8):5156–5186
18. Sahariah P, Másson M (2017) Antimicrobial chitosan and chitosan derivatives: a review of the structure–activity relationship. Biomacromol 18(11):3846–3868
19. Petroni S et al (2023) Chitosan-based biomaterials: insights into chemistry, properties, devices, and their biomedical applications. Mar Drugs 21(3):147
20. Ahmed S, Ikram S (2015) Chitosan & its derivatives: a review in recent innovations. Int J Pharm Sci Res 6(1):14–30
21. Zargar V, Asghari M, Dashti A (2015) A review on chitin and chitosan polymers: structure, chemistry, solubility, derivatives, and applications. ChemBioEng reviews 2(3):204–226
22. Pillai CK, Paul W, Sharma CP (2009) Chitin and chitosan polymers: chemistry, solubility and fiber formation. Prog Polym Sci 34(7):641–678
23. de Moura CM et al (2011) Evaluation of molar weight and deacetylation degree of chitosan during chitin deacetylation reaction: used to produce biofilm. Chem Eng Process 50(4):351–355
24. Kadyseva OV et al (2021) Study of the effect of the physicochemical properties of chitosan on its haemostatic activity. Progr Chem Appl Chitin Its Deriv 26:112–120
25. Chen C-Y et al (2008) Characteristics of molybdate-impregnated chitosan beads (MICB) in terms of arsenic removal from water and the application of a MICB-packed column to remove arsenic from wastewater. Biores Technol 99(16):7487–7494
26. Wani TU, Pandith AH, Sheikh FA (2021) Polyelectrolytic nature of chitosan: influence on physicochemical properties and synthesis of nanoparticles. J Drug Deliv Sci Technol 65:102730
27. Kasaai MR (2007) Calculation of Mark–Houwink–Sakurada (MHS) equation viscometric constants for chitosan in any solvent–temperature system using experimental reported viscometric constants data. Carbohyd Polym 68(3):477–488
28. Khurana SK et al (2021) Bovine brucellosis—a comprehensive review. Vet Q 41(1):61–88
29. Alavi M, Nokhodchi A (2023) Micro-and nanoformulations of antibiotics against *Brucella*. Drug Discov Today 28(12):103809
30. Razei A et al (2017) Application of nanoparticles drug delivery systems in the treatment of intracellular bacterial infections. Minerva Biotecnologica 29(3):156–165
31. Díaz AG et al (2016) Spray dried microspheres based on chitosan: a promising new carrier for intranasal administration of polymeric antigen BLSOmp31 for prevention of ovine brucellosis. Mater Sci Eng, C 62:489–496
32. Sadeghi Z et al (2020) Mannosylated chitosan nanoparticles loaded with FliC antigen as a novel vaccine candidate against *Brucella melitensis* and *Brucella abortus* infection. J Biotechnol 310:89–96
33. Petrovici AR et al (2023) Dextran-chitosan composites: antioxidant and anti-inflammatory properties. Polymers 15(9):1980
34. Divya K, Smitha V, Jisha M (2018) Antifungal, antioxidant and cytotoxic activities of chitosan nanoparticles and its use as an edible coating on vegetables. Int J Biol Macromol 114:572–577
35. Hussein AA, Aldujaili NH (2020) Antimicrobial, antibiofilm, and antioxidant activity of chitosan nanoparticles synthesized by *E. coli*. J Phys Conf Ser
36. Vinsova J, Vavrikova E (2011) Chitosan derivatives with antimicrobial, antitumour and antioxidant activities-a review. Curr Pharm Des 17(32):3596–3607

37. Kim S (2018) Competitive biological activities of chitosan and its derivatives: antimicrobial, antioxidant, anticancer, and anti-inflammatory activities. Int J Polym Sci 2018(1):1708172
38. Nguyen HT et al (2023) Synthesis, properties, and applications of chitosan hydrogels as anti-inflammatory drug delivery system. J Porous Mater 30(2):655–670
39. Ayaz F, Demir D, Bölgen N (2021) Differential anti-inflammatory properties of chitosan-based cryogel scaffolds depending on chitosan/gelatin ratio. Artif Cells Nanomed Biotechnol 49(1):682–690
40. Azuma K et al (2015) Anticancer and anti-inflammatory properties of chitin and chitosan oligosaccharides. J Funct Biomater 6(1):33–49

# Chapter 2
# Synthesis Protocol

**Abstract** The synthesis of CSNPs comprises a diverse array of fabrication strategies that enable tailored control over morphology, particle size, loading capacity, and overall functionality. A wide range of synthesis techniques are presented to synthesize CSNPs such as reverse micelles, ionotropic gelation, top-down approaches, polyelectrolyte complexation, desolvation, nanoprecipitation, spray-drying, covalent cross-linking, incorporation or incubation-based formulations, complex coacervation, emulsion droplet coalescence, and emulsion solvent diffusion procedures. These methodologies further allow the creation of unique nanoarchitectures such as drug-polymer conjugates, metal-functionalized hybrids as well as hydrogels and core-shell systems. Representative applications including alginate and neem oil encapsulated CS nano-formulations, curcumin loaded CS-gum arabic complexes, PEGylated CSNPs as promising carriers, zein–honey–CS composites, CS-curcumin gold nanoconjugates, bovine serum albumin-loaded CSNPs, and gadopentetic acid-encapsulated CSNPs highlight the versatility of CS as a customizable platform for pharmaceutical, biomedical, and environmental applications. The synthesis routes outlined demonstrate the significant versatility of CS nanomaterials, highlighting their potential for targeted delivery, enhanced bioactivity, and multifunctional therapeutic applications.

**Keywords** CSNPs · Biomedical · Nanomaterials · PEGylated · Top-down

In this chapter, the most recent and thorough synthesis procedure that is involved in the fabrication of CSNPs will be discussed.

## 2.1 Reverse Micelles Method

Currently, microemulsion systems (MEs) are renowned as size-controllable nanomaterial synthesis techniques. When surfactant concentration exceeds critical micelle concentration, reverse micelles form in water-in-oil microemulsion systems (W/O

 
A. Haider et al., *Chitosan Nanoparticles for Brucellosis Treatment*, Springer Series in Biomaterials Science and Engineering 21, https://doi.org/10.1007/978-3-032-16292-2_2

MEs) (Fig. 2.1) [1]. Reverse micelles are stable liquid solutions formed by water, surfactant, and oil [2]. A wide variety of surfactants can form reverse micelles, including cationic, anionic, nonionic, as well as zwitterionic molecules. Among anionic surfactants which form reverse micelles in diverse solvents, the most recognized is AOT or sodium 1,4-bis-2-ethylhexylsulfosuccinate, which forms spherical reverse micelles in aliphatic and aromatic solvents excluding cosurfactant. Remarkably, the variable water amount that might be solubilized varies depending on several parameters such as temperature and external solvent [3].

Chitosan nanoparticles (CSNPs) can also be synthesized by the reverse micelles method via utilizing heptane as solvent along with hexamine as surfactant. Particle size of CS nanoparticles can be controlled in this method. Water combined with organic solvents utilizing surfactants to create reverse micelles. Surfactants encapsulate water within solvents and are incorporated in synthesis to diminish crystal size. Their varied alkyl-chain lengths influence the morphology, size, and crystalline structure of the material. A solution of chitosan was prepared by dissolving it in 0.1 M acetic acid, resulting in a clear liquid. Additionally, 2 g of hexamine were dissolved in 10 ml of heptane. The two solutions were combined and subjected to a magnetic stirrer for a duration of 24 h. The interaction between the two solutions results in the formation of CSNPs, which can subsequently be dried in a microwave oven. The molecular structure of CS undergoes modifications throughout the process [5].

Substantial advantage of the reverse micelle approach is scalability and microfluidic systems for synthesis to produce greater quantities of catalysts [6]. ME systems

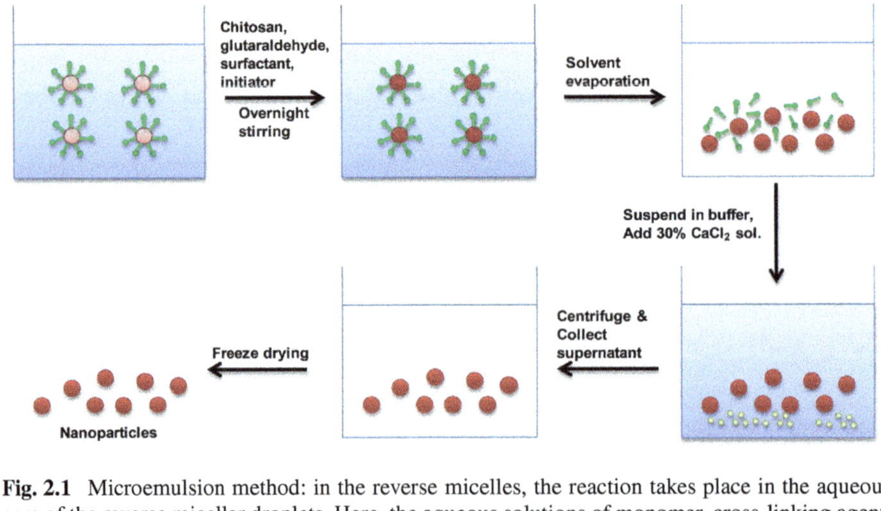


**Fig. 2.1** Microemulsion method: in the reverse micelles, the reaction takes place in the aqueous core of the reverse micellar droplets. Here, the aqueous solutions of monomer, cross-linking agent, and other hydrophilic compounds remain in the aqueous core (host nanoreactor) of the reverse micelles. Polymerization reaction, which leads to the formation of the nanoparticles, takes place within these aqueous cores by a primary growth process. Reproduced with permission from Ref. [4]

have the capability to stabilize NPs without any protecting agent in addition to optimistic size-controlling proficiency. Thus, inherent advantages of ME systems are of great attention to nano-scientist since MEs are appropriate templates for producing monodisperse and stable nanomaterials mainly for specific precious metals that destabilize or aggregate rapidly in colloidal form, as observed with reported colloidal particle systems based on rhenium [1].

## 2.2 Ionotropic Gelation (IG)

Ionotropic gelation has been used in polymeric nano- as well as microparticles synthesis for biomedical applications since the 1990s. CS and sodium tripolyphosphate (STPP) are well explored and mostly employed ionic pairs for this method (Fig. 2.2). Biologically active molecules are incorporated into reactions, becoming ensnared within the physical network established by ionic species in drug delivery methodologies. Additionally, various charged groups play a significant role; for example, the advantages of protein encapsulation arise from the presence of both positive and negative charges in amino acids, which facilitate binding with counterions. Examine the model system comprising chitosan (CS) and sodium tripolyphosphate (STPP). Basic amino acids, which possess an additional amino group, engage in interactions with the highly electronegative phosphate groups of STPP. Conversely, acidic amino acids are drawn to the positively charged chains of CS, facilitating the encapsulation of interferon alpha, insulin, bovine serum albumin (BSA), and cationic peptides, achieving remarkable encapsulation efficacy within CS-TPP NPs. A multitude of studies have been undertaken to improve drug efficacy by prolonging bioavailability and stability, revealing the promising role of IG in the field of biomedicine. Cellular uptake of molecules encapsulated within 200 nm CS NPs was observed after 24 h, particularly in the context of the anticancer agent resveratrol and BSA. The ionotropic gelation of chitosan can be executed using a range of anionic agents, notably in conjunction with alginate [7].

### *2.2.1 CS-Alginate Nanoparticles*

CS-alginate beads are generated through electrostatic interactions between chitosan (CS) and alginate. Particles derived from IG possess a CS core and a CS-alginate surface [9]. IG enhances particle stability by incorporating an additional polymeric protective layer. The standard procedure is modified through pre-gelation, where one polymer interacts with a metal salt, a small ionic agent, followed by polyelectrolyte complexation with the addition of the remaining polymer. Physically cross-linked chitosan-alginate NPs were synthesized for the encapsulation of various bioactive molecules, including insulin [7].

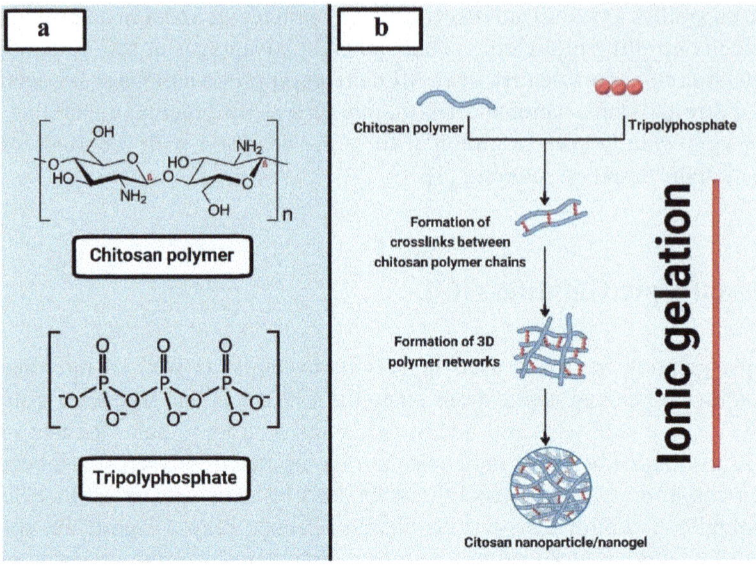


**Fig. 2.2** A molecular structure of Cs polymer and tripolyphosphate (**a**) and schematic illustration of the ionic gelation method regarding Cs NPs formation (**b**). Cs polymer contains $NH_2$ groups, which cause a positive charge, while TPP, due to its phosphate groups, is an anionic compound. The addition of TPP to Cs solution causes the formation of cross-links between the $NH_2$ groups of the Cs polymer, which leads to Cs NPs formation. Reproduced with permission from Ref. [8]

### 2.2.2 *Neem Oil-Encapsulated Chitosan (CS/NO NPs)*

CS/NO NPs prepared using ultrasound-assisted IG based on interaction between CS amino group and STPP polyphosphate group. At first, 0.45 g CS dissolved in 150 ml 1% acetic acid solution and stirred for 1 h with magnetic stirrer. Solution was filtered to eliminate immiscible impurities. Then, sodium hydroxide 2 M solution added gradually till pH reaches 4. Tween 80 added with 100 μL volume to above solution and stirred for 10 min. Neem oil homogenous solution in 4 ml butanol with varying mass ratios of CS and NO (1:0, 1:0.1, 1:0.25, 1:0.5, 1:0.75, and 1:1 g/g) added to mixture and stirred continuously for 10 min. 0.18 g of STPP was freshly prepared and added drop by drop to 40 ml of distilled water while continuously stirring for a duration of 90 min. Ultimately, the solution was treated for ultrasonic durations of 5, 10, 15, 20, and 25 min [10].

### 2.2.3 *Benefits and Drawbacks*

The IG procedure offers several additional benefits beyond notable improvements in drug encapsulation for enhanced bioavailability. The ideal interactions between polymer and drug result in 100% encapsulation efficiency. The IG method is cost-effective, requiring neither expensive reagents nor specialized equipment, and can be completed in under 10 h. The synthesis conditions can be modified to produce microparticles or NPs by altering the reagent ratios and incorporating a stabilizer. The limitation of the IG approach is a small number of studies involving various polymers, which suggests the potential for discovering new alternatives, ionic pairs, and novel reactions with significant outcomes. An additional drawback is that particle production consistently occurs in aqueous dispersion, especially when subsequent drying methods are not recommended. This process necessitates specific critical requirements, such as maintaining a constant speed for solution lowering and controlled stirring [9].

## 2.3 Top-Down Method

The top-down approach in nanofabrication involves the formation of nanostructures from larger parent superstructures through disintegration (Fig. 2.3). Implementing bottom-up strategies for synthesizing CS nanostructures is more significant than employing top-down technologies.

### *2.3.1 Ultrasonication*

CSNPs and nanofibers can be synthesized through the direct ultrasonic treatment of D-chitin nanofiber dispersion. The dispersion concentration was adjusted to 0.1%

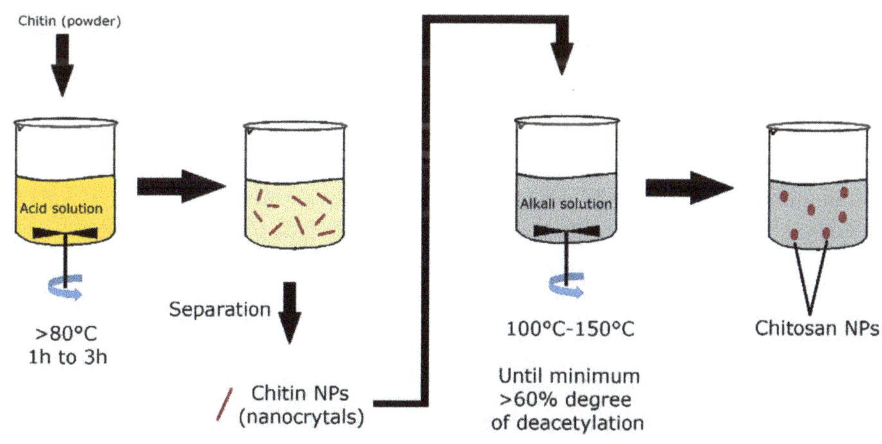


**Fig. 2.3** Schematic illustration of chitosan nanoparticle forming via top-down method. Reproduced with permission from Ref. [11]

(w/v) in a 100 ml glass beaker prior to sonication. Ultrasonication was conducted utilizing an ultrasound tip operating at a frequency of 24 kHz, delivering 360 W of power for a duration of 15 min. The centrifugation procedure was utilized for the separation of CSNPs and chitosan nanofibers (CS NFs) from water-soluble chitosan oligomers. The solution was diluted ninefold with distilled water prior to centrifugation. The solution was subsequently subjected to centrifugation at 9000 rpm to facilitate the separation of chitosan oligomers. The separation of CSNPs and CS NFs proved to be an intricate process when employing the centrifugation method, primarily due to the instability of the resulting dispersion. Nevertheless, CSNPs may be isolated through the application of gravity filtration employing ash-less cellulose Whatman filter papers, grade 42. The filtrate comprised CSNPs that are almost entirely lacking in CS NFs. The residue was primarily composed of CS nanofibers, accompanied by a few CSNPs [12].

### *2.3.2 Electrospinning*

Electrospinning is a fiber formation process wherein polymer melts or solutions are subjected to high voltage to generate fine jets. Recently, electrospun composite nanofibers have been developed using CS in conjunction with synthetic polymers such as polyvinyl alcohol (PVA), polyethylene terephthalate (PET), polylactic acid (PLA), polycaprolactone (PCL), polyethylene oxide (PEO), and nylon-6. The incorporation of these composite nanofibers offers advantages over pure chitosan electrospun nanofibers, as their inclusion significantly enhances mechanical, antimicrobial, biocompatible, and other characteristics of the nanofibers. CS/PVA nanofibers can be prepared through the electrospinning of aqueous solutions containing CS/PVA/acrylic acid, with adjustments made to the concentrations of each component. Pure CS nanoporous fibers can be obtained by dissolving PVA in CS/PVA bicomponent fibers using an aqueous NaOH solution. These pure CS nanofibers exhibit exceptional properties, including excellent biocompatibility, significant porosity, and a high surface-to-volume ratio. CS/PVA nanofibers were cross-linked using heat-induced chemical cross-linking. Triethylene glycol dimethacrylate (TEGDMA) was incorporated into the CS/PVA solution prior to spinning, followed by heat treatment of the as-spun mats for 2 h at 80 °C. Results indicated that nanofibers exhibited a smooth surface and consistent morphology, with tensile strength of the NFs improved as TEGDMA content increased. Electrospun CS NFs possess applications in wound dressing, enzyme immobilization, filtration, drug delivery, tissue engineering, and catalysis [13].

Trifluoroacetic acid (TFA) is essential for effective CS electrospinning; however, its widespread application is constrained due to significant toxicity concerns. Following the successful electrospinning of CS using TFA, various groups have reported the effective electrospinning of CS NFs from a more environmentally friendly acetic acid solution. Success resulted from a lower degree of deacetylation of CS (~ 60%) and a high concentration of acetic acid (90%). The increase in acetic acid concentration from 10 to 90% resulted in a decrease in surface tension, while viscosity remained practically unchanged. Reducing surface tension while increasing jet charge density is essential for addressing the challenges associated with CS electrospinnability. A moderate molecular weight of approximately 100 kDa is optimal for generating NFs with a smooth surface and a narrow polydispersity, resulting in an average diameter of ~ 130 nm. Conversely, low molecular weight CS yielded fragile fibers with bead-like structures, whereas high molecular weight CS resulted in nanofibers characterized by a rough surface and significant diameter polydispersity. Electrospinning of CS was successfully achieved using 5% acetic acid in conjunction with small quantities of 1-ethyl-3-(3-dimethylaminopropyl) carbodiimide hydrochloride (EDC) at a ratio of CS to EDC of 30:1. The fibers exhibited a diameter range of 53 ± 20 nm and demonstrated significant antioxidant and antibacterial characteristics. This procedure was effective solely for CS with a high molecular weight of 310 kDa [14].

## 2.4 Polyelectrolyte Complex Method

The development of colloidal polyelectrolyte complexes (PECs) represents a safe and environmentally friendly method for fabricating materials intended for pharmaceutical applications. PECs are formed in water without the inclusion of chemical surfactants, cross-linkers, or organic solvents, through the spontaneous ionic condensation of polyelectrolyte solutions with opposite charges (Fig. 2.4). The fabrication method is facile to implement through one-shot addition and is energy efficient, as it operates at low shear rate at standard room temperature [15]. PEC NPs have garnered increasing interest due to their potential applications in micro-encapsulation and the controlled release of drugs, proteins, enzymes, and DNA. These systems are suitable for loading a limited number of charged drug molecules, offering several advantages, including facile preparation, favorable structural stability, low toxicity, high drug loading capacity, prolonged circulation in the bloodstream, and controlled, targeted delivery that can be modulated by physical or chemical stimuli like pH and ionic strength [16].

The PEC technique entails the formation of CSNPs through the addition of counter ions or oppositely charged polymers to chitosan in solution. Counter ions or oppositely charged polymers are dissolved with CS in an acetic acid solution while stirring under standard conditions. Electrostatic interactions between oppositely charged colloidal systems and additional polymers or counter ions lead to charge neutralization, resulting in the formulation of polyelectrolyte complexes. Consequently, charge neutralization in PEC self-assembly leads to a significant reduction in hydrophilicity. NPs produced using this method are reported to range in size from 50 to 700 nm. PEC-complexed CSNPs serve as delivery systems for proteins, drugs, plasmid DNA, and peptides (Fig. 2.5). Heparin-loaded CSNPs can also be produced via this method. The molecular weight, pH of the solution, and concentration of constituents significantly influence the yield size of NPs. Decreased pH levels and intermediate molecular

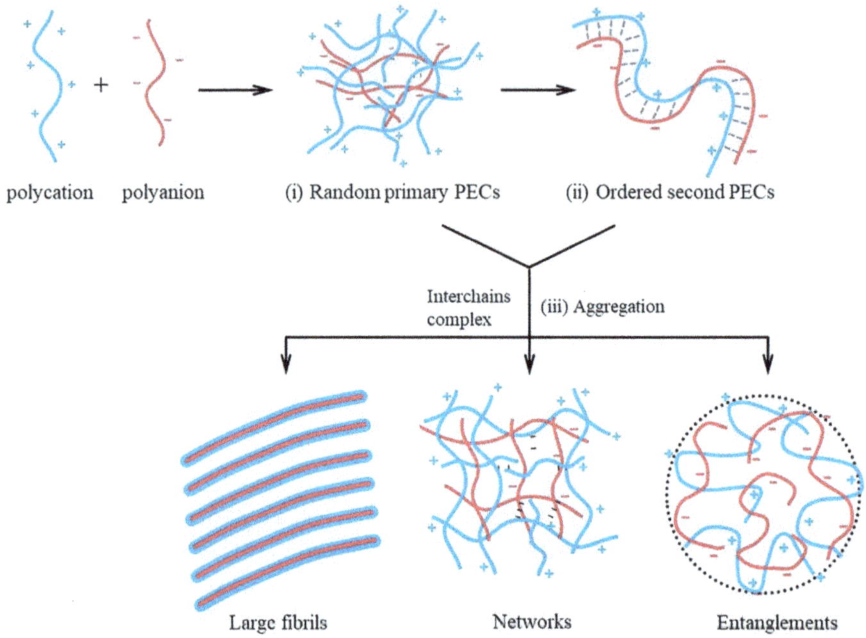


**Fig. 2.4** Scheme of the different steps of the polyelectrolyte complexes formation. Reproduced with permission from Ref. [15]. Copyright 2020, Elsevier

weight values result in enhanced NP complexation. The synthesis of IgA-loaded CS-dextran NPs via the PEC process offers a straightforward and efficient method for developing drug delivery systems. CSNPs synthesized by the authors exhibited size ranges of 300–500 nm, with Zeta potential (ζ-potential) values ranging from + 40 to + 50 mV [17].

Gum arabic (GA) is a biodegradable and biocompatible polysaccharide, extensively utilized as an emulsifying and suspending agent in topical, oral pharmaceutical, and food formulations. Carboxyl groups of GA mostly dissociate around neutral pH, resulting in a highly charged, open, and expanded molecular structure. This attribute contributes to strong surface activity and the ability to form viscoelastic films. Furthermore, GA exhibits a greater number of interaction sites and a negative charge conducive to binding with polycationic CS. Therefore, the combination of GA and CS is effective in the formation of PEC. Electrostatic complexes between GA and CS are presumed to create durable viscoelastic films encasing oil droplets, providing a physicochemical barrier against oxidation.

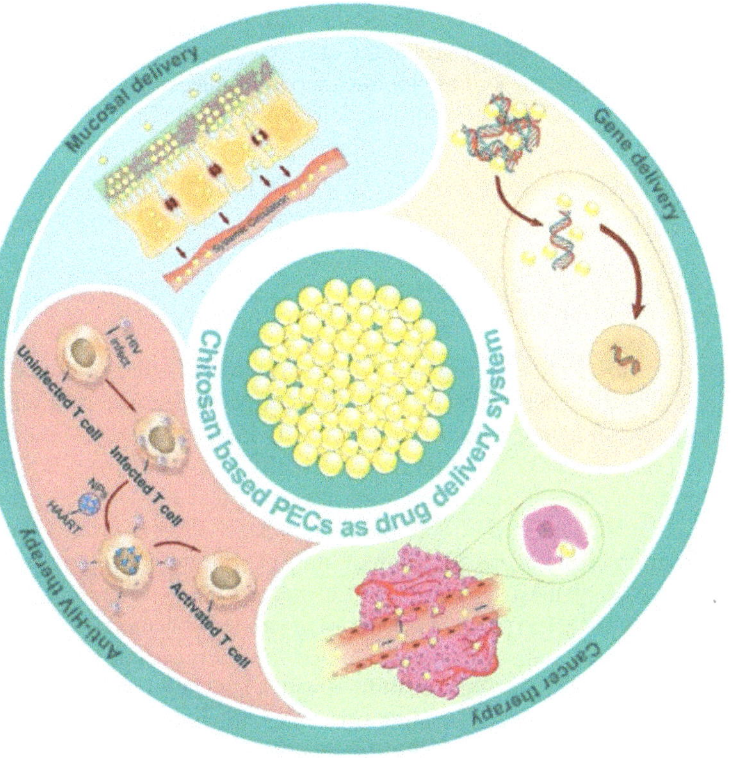


**Fig. 2.5** Schematic illustration of chitosan-based PECs for various biomedical applications. Reproduced with permission from Ref. [15]. Copyright 2020, Elsevier

## *2.4.1 Curcumin-Loaded Chitosan-Gum Arabic NPs*

Curcumin-loaded CS-GA NPs were synthesized by dissolving CS in 1% (v/v) acetic acid solution under vigorous stirring until fully dissolved. The solution was subjected to filtration using a filter with a pore size of 30–50 μm to eliminate any insoluble substances. In order to formulate the GA solution, gum arabic was solubilized in distilled water, ensuring that the pH was aligned with that of the CS solution. Thereafter, the GA solution was incrementally introduced to the CS solution in equal volumes over a duration of 30 min, utilizing an IKA RW20 digital stirrer set at 800 r/min. Curcumin-loaded CS-GA NPs were synthesized through the ethanol injection method. Initially, curcumin is dissolved in ethanol, followed by the incorporation of Tween 80 and EYPC in a specific weight ratio. The formulations contained initial curcumin concentrations of 2%, 4%, 6%, 8%, and 10%, respectively. Subsequent to dissolution, the mixture was dispersed into a chitosan solution, followed by the addition of GA, as delineated in the preparation of CS-GA NPs. The system was subsequently connected to the rotary evaporator for the purpose of ethanol removal.

Final samples were meticulously placed in vials beneath a nitrogen atmosphere and subsequently stored in a refrigerator at 4 °C, shielded from light until their utilization. A resilient matrix of biopolymer networks encasing curcumin-emulsifier droplets could potentially improve the delayed release of curcumin within a simulated gastrointestinal tract (GIT). The synthesized CS-GA NPs demonstrate the capability to stabilize, encapsulate, and effectively deliver curcumin [18].

### *2.4.2 PEGylated Chitosan NPs*

PEGylated stealth NPs have emerged as promising carriers for drug delivery in cancer treatment. CS is characterized by its biocompatibility, biodegradability, nontoxicity, and nonimmunogenic properties. It is grafted with PEG to improve solubility while maintaining biocompatibility. Initially, the Schieffler base was formed through the reaction of formaldehyde with the amino group of CS, followed by the grafting of methoxypolyethylene glycol onto chitosan. CS-g-mPEG is formed through the addition reaction of the hydroxyl group on mPEG with the Schieffler base. PEGylated CSNPs were synthesized using the PEC method. CS-g-mPEG and polyanionic poly(NIPAAm-co-AMPS) (PNAM) solutions of varying concentrations were filtered through a G3 funnel and subsequently diluted to different concentrations accordingly. A specific volume of PNAM solution was added at a rate of 2 μL/s into the CS-g-mPEG solution while maintaining continuous stirring for approximately 2–3 h. Dispersion was filtered using a G4 filter funnel in conjunction with a 450 μm filter membrane to obtain PEGylated CSNPs (CS-g-mPEG/PNAM), where the polymer functions as the nucleus and CS-g-mPEG acts as the shell. The synthesized water dispersion system was centrifuged at 13,200 rpm for 40 min at 4 °C, followed by washing with a 1% HAc solution. Precipitation was freeze-dried at − 46 °C to obtain the product following centrifugation. The results indicate that the cellular uptake competence of PEGylated CSNPs is four times lower than that of conventional CSNPs. This suggests that PEGylated CSNPs significantly reduce plasma protein adsorption, thereby suppressing recognition and macrophage phagocytosis, which extends their circulation time in the bloodstream [19].

## 2.5 Desolvation

In previous decades, the desolvation technique was successfully implemented for the generation of protein NPs. A simple desolvation procedure was employed for the synthesis of CSNPs to enhance formation yield, hydrophobic drug encapsulation capacity, and physical stability [20]. Recently, there has been growing interest in Quality by Design (QbD) as a systematic approach to ensure product quality, aiming to achieve objectives such as cost reduction, logical controls, faster time-to-market, and enhanced efficiency. CS-loaded Vildagliptin NPs were prepared using an intuitive desolvation method incorporating the QbD concept. Vildagliptin-loaded chitosan NPs were developed using a standard desolvation method with minor modifications. In 50 ml of a 2% aqueous acetic acid solution, chitosan was dissolved at varying concentrations of 200 mg, 400 mg, and 600 mg, respectively, prior to the addition of 100 mg of VLD to create a homogeneous solution. Desolvation was achieved by the gradual addition of acetone while maintaining continuous stirring until a stable low turbidity reached. Subsequently, varying volumes of glutaraldehyde (GA) were utilized as a cross-linking agent, specifically 50 μl, 100 μl, and 150 μl. The solution was agitated overnight at 600 rpm using a mechanical stirrer to facilitate acetone evaporation. NPs were obtained through centrifugation at 20,000 rpm and 4 °C for a duration of 1 h. Triplicate sediment was washed under identical conditions. CS-VLD NPs demonstrate improved therapeutic efficacy and prolonged drug activity, thereby improving patient outcomes and treatment adherence in Type 2 Diabetes Mellitus relative to conventional drug delivery methods [21].

The desolvation process can also be employed to produce micron-sized carrier molecules. CS NPs are often synthesized using a process derived from sodium sulfate. This method enables the distribution of protein and DNA. Coacervation and phase separation serve as fundamental principles of the desolvation technique. NPs are precipitated through the addition of desolvation agents such as acetone or ethanol, while stability is achieved by incorporating a cross-linking agent. This method involves treating an aqueous CS solution containing a stabilizing agent, Tween 20, with a precipitation agent such as sodium sulfate. Water-encircling CS is effectively eliminated by a saline CS solution. CS's insolubility results in precipitation. Glutaraldehyde was utilized to harden NPs with an average size of 373 ± 71 nm. This procedure offers several advantages over alternative methods, notably its capacity to produce NPs in a single step, as well as its low cost, reduced electricity consumption, and frequency [22].

## 2.6 Nanoprecipitation

Nanoprecipitation is also referred to as the solvent-displacement method and provides several advantages over alternative techniques. It is swift and simple to carry out. NP creation is instantaneous, and the procedure occurs in a single-step approach. Complex hydrodynamic and phase-separation processes are involved in nanoprecipitation. The mechanism of NP formation is described in the context of interphase turbulence between two liquid phases, involving diffusion, flow processes, and surface interactions at a non-equilibrium stage. The development of submicron particles is influenced by the combination of operating conditions, the composition of organic and aqueous phases, and the physicochemical interactions between these phases [23].

### 2.6.1 Zein-Based Honey and Chitosan NPs via Flash Nanoprecipitation

Flash nanoprecipitation (FNP) procedure is an emerging method for rapid, simple, and polymer-directed NPs self-assembly for effective encapsulation of functional agents. FNP exhibits several advantages such as extraordinary loading efficacy (typically > 80%), fast production time within milliseconds, limited size distribution, as well as easy scale-up. Water-insoluble prolamin found from corn is zein and generally acknowledged as safe by Food and Drug Administration (FDA) with wide-ranging applications. Chitosan stabilizes zein NPs by decreasing aggregation and sedimentation through increasing electrostatic repulsion among particles. Versatile polymer CS is also recognized as safe by FDA. To prepare zein-honey-chitosan NPs, zein (2.5 mg/mL) and CS (10 mg/mL) are dissolved in ethanol solution (80% v/v) and aqueous solution of glacial acetic acid (1% v/v) correspondingly. Honey was either dissolved in water or in a zein solution at a concentration of 10 mg/mL. Four experiments were conducted in triplicate using an ethanolic stream consisting of zein as one component and either water or a CS solution as the second component. The initial experiment involved mixing a zein solution (2.5 mL) with an equal volume of $H_2O$ to assess the characteristics of empty NPs. Honey-loaded NPs were produced by dissolving honey in either a zein solution or water. Chitosan was incorporated to stabilize honey-loaded zein NPs. The resulting solution from the collision of two streams was collected in 45 mL of Milli-Q water, achieving a final volume of 50 mL. The NP formulation (zein-honey-chitosan) was evaluated at pH levels of 3, 7, and 10, with size distribution and ζ-potential assessed to determine stability over time. Vitamin C was utilized as a hydrophilic compound model, encapsulated in NPs and dissolved in an 80% zein-honey ethanolic solution (Fig. 2.6). The produced NPs exhibited appropriate size, surface charge, and polydispersity index (PDI) for biomedical applications [24].

### 2.6.2 Chitosan-Curcumin Conjugated Gold Nanoparticles

In another research, CS containing highly loaded curcumin-conjugated gold NPs produced employing the antisolvent principle via FNP by vortex mixer with four inlets linked to four hermetic stainless syringes through Teflon tubing (Fig. 2.7). First syringe contained 50 mL aqueous solution of surfactant Tween 80 (0.5 M), the second was hydrogen tetrachloroaurate ($HAuCl_4$) (0.01 M) aqueous solution with yellow color and surfactant Tween 80 (0.5 M), the third was NaOH aqueous solution to adjust pH at 9.3, and the fourth was curcumin with CS in tetrahydrofuran (THF) organic solvent at 5 mg/mL concentration with yellow color. Syringe pumps were employed digitally to regulate the injection speed ratio of fluid streams. After mixing, progressive and rapid change in color occurred from pale yellow to turbid/burgundy red confirming gold NPs formation. Control sample prepared without $HAuCl_4$ aqueous

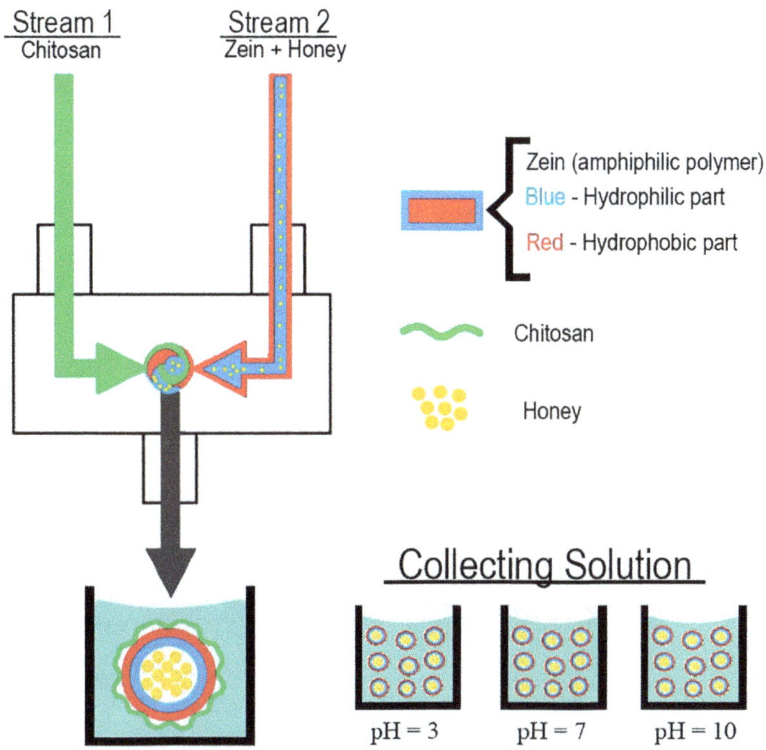


**Fig. 2.6** The proposed approach for zein–honey–chitosan NPs synthesis by FNP. Reproduced with permission from Ref. [24]

solution and water substituted with this solution to synthesize CS containing highly loaded curcumin for comparative analysis of antibacterial test.

Prepared CS with highly loaded curcumin-conjugated gold NPs as compared to pure CS and curcumin-encapsulated CSNPs with high antibacterial action (Fig. 2.8) [25].

## 2.7 Spray Drying

Spray drying is extensively utilized in the production of NPs and microparticles for the pharmaceutical and food industries due to its cost-effectiveness and ease of scalability relative to alternative methods. The procedure is adaptable to commonly used processing equipment and produces particles with exceptional properties (Fig. 2.9). The microencapsulation of antibiotics such as vancomycin, amoxicillin, and ampicillin using chitosan through a spray drying process is widely employed in industrial applications to improve antibacterial efficacy. The spray-drying method yields fine

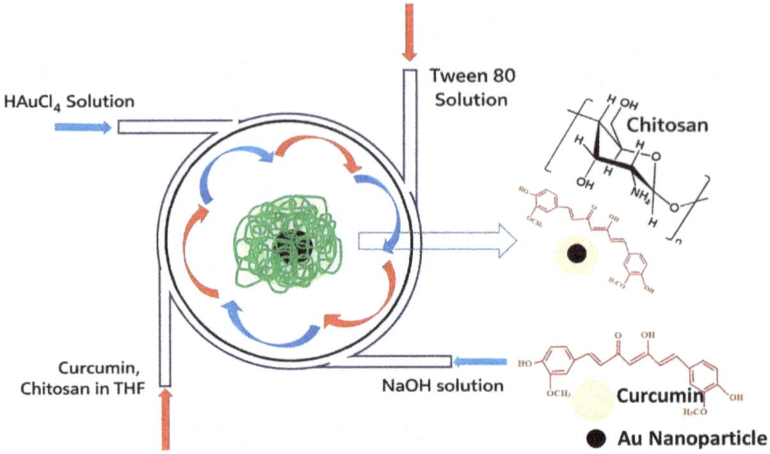


**Fig. 2.7** The flash nanoprecipitation through multi-inlet vortex mixer containing four inlets to synthesize chitosan-containing high loading of curcumin-conjugated gold nanoparticles. Reproduced with permission from Ref. [25]

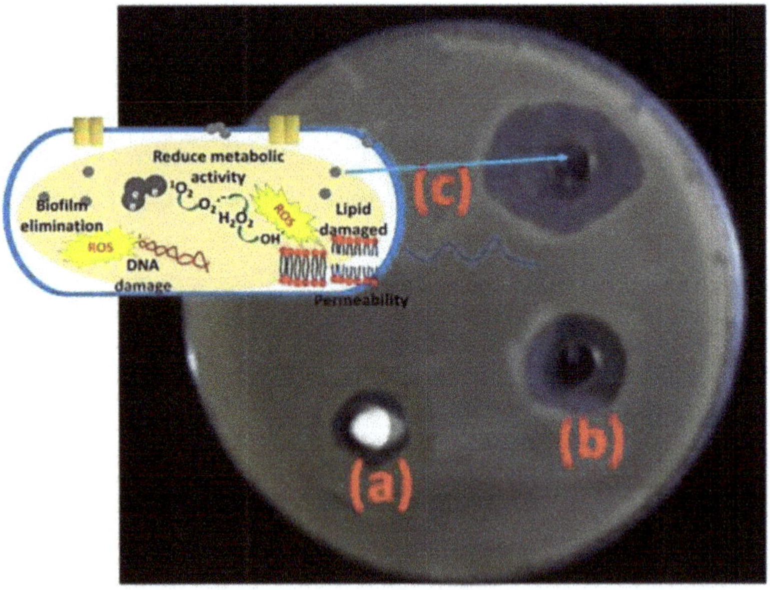


**Fig. 2.8** Antibacterial effect of **a** pure chitosan, **b** chitosan-containing high loading of curcumin, and **c** synergistic effect and oxidative stress mediated by chitosan-containing high loading of curcumin-conjugated gold nanoparticles. Reproduced with permission from Ref. [25]

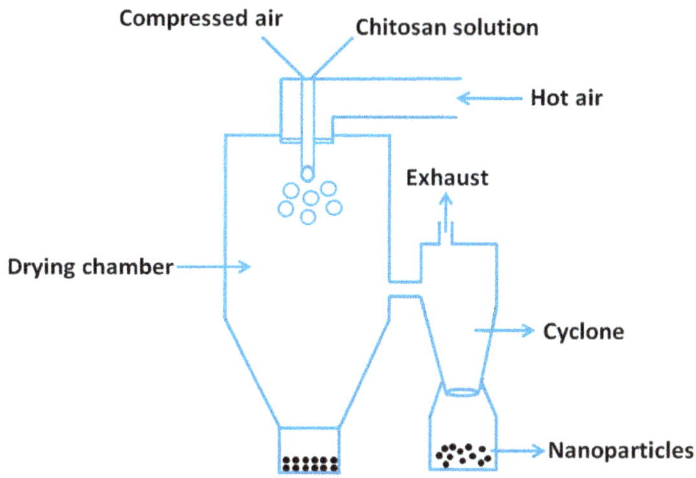


**Fig. 2.9** Preparation of chitosan particulate systems by spray-drying method. Reproduced with permission from Ref. [27]

chitosan particles ranging from 10 to 50 μm, thereby enhancing direct antibacterial action [26]. Drug-loaded agglomerates, granules, or powders can be produced using the spray-drying technique. A mixture of excipient and drug is atomized into fine droplets and subsequently introduced into a chamber filled with hot air. Drug-loaded particles were obtained as droplets dried [27].

### *2.7.1 Fabrication of CSNPs via TPP Ionic Gelation Combination with Spray Drying*

Chitosan with several molecular weights of 130, 276, 760, and 1200 cPs was dissolved in acetic acid 0.35% (w/v) solution and kept for whole night at standard room temperature. CS solution was adjusted to pH ~ 5.5 through 0.5 M NaOH solutions. TPP solution 0.25% (w/v) was dissolved in deionized (DI) water. Then, CSNPs were prepared via dropping 0.25% TPP solution into CS solution at 6:1 (w/w) of CS:TPP ratio while stirring at 900 rpm with a magnetic stirrer at room temperature for 60 min. Suspensions were centrifuged at 5000 rpm for removal of precipitants, and after that, CSNPs were collected by nanospray dryer, placed in a sterile bench room with different sized nozzles of spray dryer as 4.0, 5.5, 7.0 μm; flow rate was 120 ml/min, 1.3 $m^3$/min was drying gas flow, inlet temperature recorded was 120 °C while 80 °C was outlet temperature. CSNPs were placed inside glass tubes and stored at 5 °C.

Size and ζ-potential value of CS nanoparticles fabricated via TPP ionic gelation technique in combination with spray drying relied clearly on molecular weight.

CSNPs can strongly inhibit *Streptococcus pneumoniae* growth at 60 μg/mL concentration. Antibacterial potential of CSNP and amoxicillin complex is threefold stronger than amoxicillin alone [28].

### 2.7.2 Synthesis of CS-Amorphous Calcium Phosphate Nanocomposites

CS-amorphous calcium phosphate (CaP) nanocomposite can be synthesized via co-precipitation induced spray drying technique. Two solutions with different organic or inorganic mass ratios were prepared by dissolving 8.05 g and 0.80 g of CS for the CS/ACP.8.0 and CS/ACP.0.8 samples, respectively, along with 5 g of mono-hydrate calcium acetate (CA) in 1 L of 0.33% (w/w) acetic acid solution. Phosphoric acid (PA) 1.19 mL volume was added drop by drop in less than 60 s in above solutions under robust stirring to prevent CaP local saturation. Both solutions contained pH values of 4.6 and 5.0 before adding PA, which became 4.3 and 4.9 after PA addition. After PA addition, solutions were instantly spray dried by utilizing Mini Spray Dryer B 290 with inset temperature 160 °C, air flow capacity of 414 L/h, aspiration rate 90%, along with liquid feed rate 0.3 L/h. Both conditions had an outlet temperature below 100 °C to avoid thermal degradation. TGA analysis was used to determine organic or inorganic ratios of spray dried products [29].

## 2.8 Covalent Cross-Linking Method

Although CS is considered a highly applicable polymer, it has several drawbacks, primarily due to its high aqueous solubility, which poses challenges in applications involving aqueous biological media. The introduction of a cross-linker can resolve this issue. Cross-linking occurs via chemical reactions that connect two polymers through ionic or covalent bonds, as well as through weaker bonding interactions, for example, van der Waals forces. The formation of interpenetrating polymer network (IPN) is a significant area of research regarding cross-linking. Cross-linkers may exist as oxides, metals, amino acids, and polymers, among other chemical entities [30]. The presence of reactive groups in CS renders it amenable to modification for network generation via covalent bonding with cross-linking agents. The chemical cross-linking of CS chains facilitates the free diffusion of bioactive materials or water, enhancing mechanical properties. The most utilized cross-linking agents for CS include epoxides and aldehydes, with particular emphasis on glutaraldehyde. Chitosan cross-linking, when combined with various procedures that limit particle size to colloidal dimensions, can be utilized to create hydrogels at both nanometric and microscopic scales [31].

### 2.8.1 *Genipin-Cross-Linked Chitosan Hydrogels*

CS is effectively cross-linked using genipin, with glycerol serving as a plasticizer. CS was solubilized by stirring for 3 h at concentrations of 1% and 2% (v/v) in acetic acid to obtain solutions of 2% and 6% (w/w), correspondingly. This duration was selected to facilitate effective CS solubilization while minimizing solvent evaporation. Solutions were stored in the refrigerator for 72 h prior to use. Genipin, solubilized in 6% (w/w) ethanol, was incorporated into CS solutions while stirring, utilizing various weight ratios of genipin to chitosan. The cross-linking of CS with genipin leads to the formation of a dark-blue hydrogel [32].

### 2.8.2 *Glyceraldehyde-Cross-Linked Chitosan Hydrogels*

Glyceraldehyde (GCA) covalently cross-links with the main amino acid group present on biopolymers such as CS to form hydrogels. Genipin is generally considered less cytotoxic than common cross-linking agents which are employed for biopolymers comprising residues with primary amine groups. However, in a latest study of different cross-linker agents on cytotoxicity of four distinct cell lines, it was discovered that GCA is less cytotoxic than genipin. CS stock solution was prepared via dissolving CS in 1% (v/v) acetic acid under stirring with a magnetic stirrer to homogenize the solution for 10–12 h at ambient temperature. The pH of the solution was adjusted to prescribed values through addition of aqueous 10 M NaOH drop by drop to the CS solution. In cross-linker reactions, GCA was dissolved in MilliQ water, and the agent was added dropwise while magnetic stirring to the CS solution to achieve a fixed concentration of 1 wt%. Cross-linker concentrations ranged from 0.25 to 1 wt%, with pH adjusted to specified values. Following a 2-min stirring period, the reaction mixture was transferred onto the rheometer plate, and rheological experiments were initiated. Biocompatible GCA serves as a promising cross-linking agent for the production of tunable CS hydrogels, which can be systematically characterized through rheological methods [33].

## 2.9 Incorporation and Incubation Method

The incorporation of various proteins or drugs into biodegradable polymers has been recognized as an effective method for controlling the release profile of the retained substances. Incorporating these molecules into polymers can enhance bioavailability, efficiency, and targeting of cells or tissues. Biodegradable polymeric CSNPs are widely recognized as effective vehicles for the delivery of various proteins and nucleic acids among the available candidates. Bovine serum albumin (BSA) has been extensively studied as a model protein in drug delivery research. BSA was incorporated into CSNPs by dispersing it in collagen hydrogels, which were subsequently integrated into polylactide-co-glycolide scaffolds, facilitating controlled release while preserving the bioactivity of the released BSA.

### 2.9.1 BSA-Loaded CSNPs Preparation

CS was dissolved in a 1% (v/v) acetic acid solution at a concentration of 1.2 mg/mL, with the pH adjusted to 5 by the addition of 1 mol $L^{-1}$ sodium hydroxide (NaOH). CSNPs were formed by the addition of polyanionic TPP (1 mg $mL^{-1}$) to the CS solution in a 3:1 ratio, while being magnetically stirred at 1000 rpm for 10 min, followed by centrifugation at 20,000 rpm for 1 h. Supernatant was collected, and NPs were extensively rinsed with deionized water to remove residual sodium hydroxide, followed by freeze-drying for future use. BSA-loaded CSNPs were prepared using two distinct methods: encapsulation and adsorption. During the encapsulation process, 1% BSA was incorporated into the polyanionic phase, following the same methodology as in BSA non-loaded CSNPs. Conversely, 500 mg of previously prepared CSNPs were stirred in 50 mL of BSA at a concentration of 1.1 mg $mL^{-1}$ at 4 °C for 7 h. Following rinsing with deionized water, NPs were subjected to freeze-drying [34].

### 2.9.2 Synthesis of Thiolated CS-PHEMA Core–Shell NPs

CS was hydrated in 1 M HCl and subsequently dissolved by the addition of demineralized $H_2O$ to achieve a 1% (w/v) polymer solution. The pH was adjusted to 6.0 by the addition of 5 M NaOH. L-Glutathione reduced form (GSH) was added to the aforementioned solution in demineralized $H_2O$ while maintaining constant swirling. Subsequently, carbodiimide hydrochloride (EDAC) and N-hydroxysuccinimide (NHS) were introduced into the reaction mixture while stirring vigorously, achieving a final concentration of 200 mM for each compound. The pH was adjusted to 6.0 by the addition of 5 M NaOH. The reaction mixture was incubated at room temperature, ranging from 15 to 30 °C, for a duration of 15 h with continuous stirring. The resulting polymer conjugate was dialyzed in tubing to eliminate unbound reagents. Frozen solutions of aqueous polymer were lyophilized at − 50 °C under 0.01 mbar and subsequently stored at 4 °C prior to further use. The total amount of immobilized sulfur on the CS-GSH conjugate, in both reduced and oxidized forms, was quantified using a spectrophotometric process with Ellman's reagent. Medium MW demonstrated greater mucoadhesion compared to depolymerized chitosan, especially in its thiolated form. Furthermore, a decrease in chitosan MW led to a reduction in mucoadhesion in both unmodified and thiolated forms. NPs derived from thiolated chitosan with medium molecular weight exhibited enhanced mucoadhesion and improved penetration abilities [35].

## 2.10 Complex Coacervation Method

Complex coacervation is liquid/liquid phase separation of polyelectrolytes with oppositely charged macromolecules such as proteins, inorganic NPs, micelles, nucleotides, dendrimers, or other polyelectrolytes. The phase separation of macromolecules possessing opposing charges may be influenced by factors such as ionic strength (salt concentration), temperature, pH, stoichiometry, charge density, and the molecular weight of polyelectrolytes, as well as polyion concentration and chain flexibility [36]. The coacervation method has garnered attention for promoting synthesis of small as well as stable physical nanomaterials. It is a low-cost, simple, and reproducible technique that does not require organic solvents, energy, surfactants, or chemical cross-linkers. Hence, coacervates are of considerable interest as promising carriers for bioactive molecules such as proteins, drugs, and nucleic acids. Coacervates are usually produced via combining polysaccharide biopolymers like CS with polyanions such as carrageenan (Fig. 2.10) [37]. Among commercially existing biopolymers, iota-Carrageenan (ι-Carrageenan) is a biopolymer with a fine film-forming nature along with excellent mechanical stability [38]. ι-Carrageenan exhibits several properties such as antiviral, anticancer, anticoagulant activities, wound healing, bone promotion, and antimicrobial treatment.

To synthesize polymeric NPs, only dialyzed or purified polymers are used. Aqueous solutions of CS and ι-carrageenan were individually formulated at diverse concentrations within a 0.1 M acetic acid/sodium acetate buffer, adjusted to pH levels of 3.6, 4.0, and 4.4, and subsequently filtered using 0.45 mm polytetrafluoroethylene filters. Subsequently, the solutions were combined by directly introducing aliquots of ι-carrageenan solutions into cationic polyelectrolyte CS solutions, or conversely, by adding CS solutions into ι-carrageenan solutions (anionic polyelectrolyte). The volume-volume ratios of mixtures were calibrated to polyanion/polyanion or polycation/polyanion ratios of 2, 5, and 8. The mixtures were subsequently agitated using a magnetic stirrer at a speed of 750 rpm for a duration of one hour at ambient temperature. Subsequently, the mixtures, comprising nanoparticulate material in suspensions, were preserved for a duration of 12 h at a temperature of 4 °C to facilitate the precipitation of larger particles from the resultant suspensions. Subsequently, the suspensions were subjected to centrifugation at 3000 × *g* for a duration of 10 min to facilitate the removal of larger particles or precipitates, and were subsequently stored at 4 °C for future utilization [37].

## 2.11 Emulsion Droplet Coalescence

Emulsion droplet coalescence is based on emulsion cross-linking and precipitation principles. This method involves preparing two emulsions, one containing CS aqueous solution and liquid paraffin oil-based drug and another containing CS aqueous solution of NaOH in paraffin oil. The emulsions were merged through

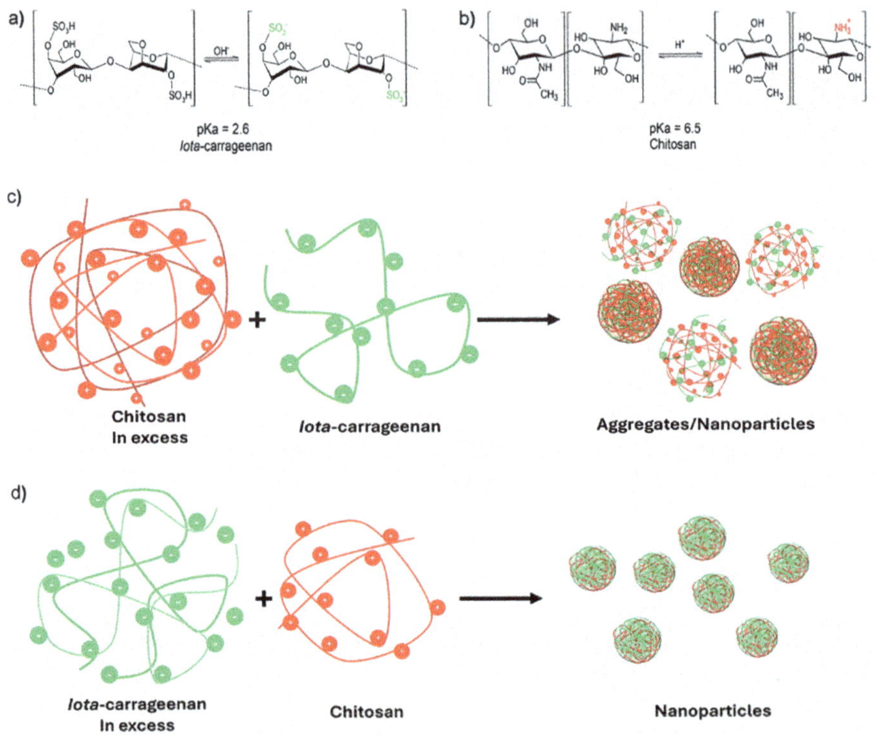


**Fig. 2.10** The chemical structures of ι-carrageenan and chitosan under equilibrium conditions in aqueous solutions are shown (**a, b**). Aggregated coacervates are formed when an excess of chitosan is used due to the partial deprotonation of chitosan chains (**c**). Conversely, smaller coacervates are obtained when an excess of ι-carrageenan is present during preparation as a result of the partial deprotonation of the sulfate groups in ι-carrageenan (**d**). Reproduced with permission from Ref. [37]

vigorous stirring, leading to the formation of droplets that may collide and coalesce, resulting in the precipitation of CS droplets and the production of small-sized particles (Fig. 2.11). This approach was employed to synthesize gadopentetic acid-loaded CSNPs [39].

### *2.11.1 Synthesis of Gadopentetic Acid-Loaded CSNPs (Gd-CSNPs)*

CS 2.5% w/v was dissolved in gadopentetic acid (Gd-DTPA) in a 5–15% w/v aqueous solution, and a 1 ml aliquot was added to 10 ml of liquid paraffin containing 5% v/v Arlacel C. A mixture was agitated to formulate a water-in-oil (w/o) emulsion A utilizing a high-speed homogenizer. In a comparable manner, w/o emulsion B, which

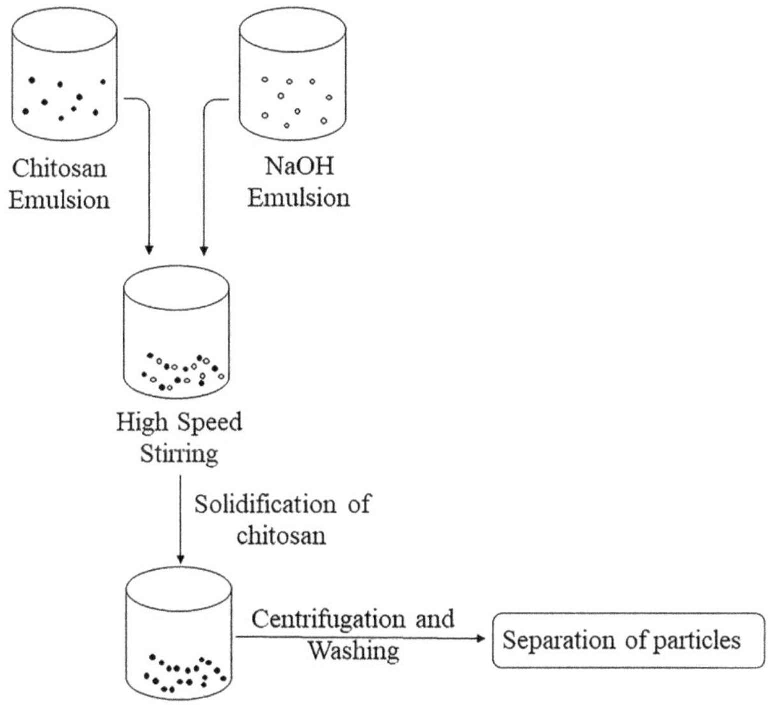


**Fig. 2.11** Diagrammatic representation of preparation of chitosan nanoparticles by emulsion droplet coalescence method. Reproduced with permission from Ref. [39]

was formulated by incorporating 1.5 ml of 3 N NaOH solution into 10 ml of liquid paraffin containing 5% (v/v) Arlacel C. Emulsion B was incorporated into emulsion A and subjected to vigorous mixing and stirring utilizing a homogenizer. Upon the coalescence of droplets, chitosan was deposited in the form of NP. The Gd-CSNPs were subjected to a mixed emulsion, followed by a process of washing and separation via centrifugation for 60 min at a speed of 3000 rpm, utilizing water, toluene, and ethanol in a sequential manner. As a result, Gd-CSNPs were acquired in the form of water suspensions and powders, which were produced through lyophilization following their dispersion in an isotonic mannitol solution [40].

## 2.12 Emulsion Solvent Diffusion

The emulsification-solvent diffusion technique is capable of producing polymeric NPs with impressive yields, eliminating the need for high shear stirring, ensuring batch-to-batch reproducibility, and achieving satisfactory encapsulation efficiencies

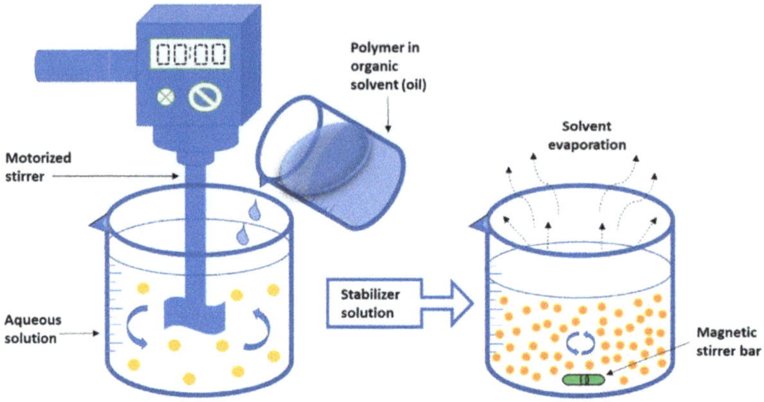


**Fig. 2.12** Diagrammatic representation of the emulsification solvent diffusion method. Reproduced with permission from Ref. [43]

(Fig. 2.12) [41]. This method offers considerable advantages compared to alternative approaches; for example, the utilization of appropriate organic solvents enables the preparation of nanocapsules and nanospheres with high yield and exceptional reproducibility, while also allowing for precise control over particle size, among other benefits. This approach consists of two distinct phases. Initially, an emulsion is formed from a blend of a partially miscible solvent in water, which has previously been saturated with water, alongside a polymer in the aqueous phase that has been saturated with the solvent and includes a stabilizer. The subsequent step following the acquisition of the emulsion involves the diffusion of solvent from the internal phase (globules) through the dilution of the emulsion with a substantial quantity of water. This alteration in composition leads to the ingress of polymer into the non-solvent phase, resulting in the formation of polymer aggregates referred to as "protonanoparticles," which possess colloidal dimensions from which NPs emerge. This process has been successfully modified, yielding significant results in the development of nano- and microspheres that contain drugs either encapsulated within a matrix or dispersed into an oil core, which is enveloped by a polymeric membrane that may be dissolved, bonded to the surface of the particles, or chemically adsorbed [42].

### *2.12.1 Synthesis of Insulin-Loaded Chitosan-Reduced Gold Nanoparticles Functionalized with PLGA (CSAuNPs/Insulin/PLGA)*

Insulin-loaded CS reduced Au NPs functionalized with PLGA prepared using the emulsion solvent diffusion technique. The initial water phase was produced by the gradual addition of 10 mL of insulin at a predetermined concentration to 10 mL of colloidal CSAu NPs solution while stirring, followed by the incorporation of 10 mL of 1% PVA solution. The second $H_2O$ phase was obtained by dissolving 0.5 g of PVA in 50 mL of distilled water at 80 °C with stirring. Oily phase generated by dissolving 100 mg of PLGA in 10 mL of ethyl acetate. Initially, $H_2O$ was added dropwise to the PLGA solution while sonication was conducted in an ice bath. Finally, 50 mL of the second water phase (PVA, 1%) was added to the first water and oil phase (previous mixture) at ambient temperature for 60 min. Organic solvents were evaporated, and the solution was ultracentrifuged for 40 min at 30,000 rpm to complete the emulsion evaporation process. NPs were extracted from the supernatant solution and subsequently redissolved in distilled water for further use. The functionalization of CSAu NPs with PLGA demonstrated significant effects in lowering blood glucose levels within 6 h following oral administration. Therefore, the modification of CSAu NPs with PLGA is a viable and effective method for the oral delivery of insulin in the treatment of diabetes [44].

## References

1. Shiri MSZ et al (2024) Synthesis study of size-controlled rhenium nanoparticle systems using reverse micelle-based methodologies. Colloids Surf, A 694:134150
2. Moza A, Thomas T, Prasad S (2024) Local drug delivery using chitosan microspheres—a review of literature. Int J Recent Adv Multidicip Res 11(04):9714–9724
3. Orellano MS et al (2017) AOT reverse micelles as versatile reaction media for chitosan nanoparticles synthesis. Carbohyd Polym 171:85–93
4. Chandra Hembram K et al (2016) Advances in preparation and characterization of chitosan nanoparticles for therapeutics. Artif Cells Nanomed Biotechnol 44(1):305–314
5. Iswarya S, Theivasanthi T, Gopinath SC (2024) Electrochemical potential of chitosan nanoparticles obtained by the reverse micellar process. Colloids Surf, A 693:134018
6. Tammaro O et al (2024) Reverse micelle strategy for effective substitutional Fe-doping in small-sized $CeO_2$ nanocrystals: assessment of adsorption and photodegradation efficiency of ibuprofen under visible light. Chem Eng J 479:147909
7. Sacco P et al (2021) Ionotropic gelation of chitosan flat structures and potential applications. Molecules 26(3):660
8. Dadashi H et al (2024) A rapid protocol for synthesis of chitosan nanoparticles with ideal physicochemical features. Heliyon 10(11)
9. Pedroso-Santana S, Fleitas-Salazar N (2020) Ionotropic gelation method in the synthesis of nanoparticles/microparticles for biomedical purposes. Polym Int 69(5):443–447
10. Pham MPT et al (2024) Ultrasound-assisted ionic gelation of neem oil-encapsulated chitosan nanoparticles as a promising biopesticide. Chem Pap 78(3):1825–1832
11. Yanat M, Schroën K (2021) Preparation methods and applications of chitosan nanoparticles; with an outlook toward reinforcement of biodegradable packaging. React Funct Polym 161:104849
12. Wijesena RN et al (2015) A method for top down preparation of chitosan nanoparticles and nanofibers. Carbohyd Polym 117:731–738
13. Sun K, Li Z (2011) Preparations, properties and applications of chitosan based nanofibers fabricated by electrospinning. Expr Polym Lett 5(4)
14. Anisiei A, Oancea F, Marin L (2023) Electrospinning of chitosan-based nanofibers: from design to prospective applications. Rev Chem Eng 39(1):31–70
15. Wu D et al (2020) Chitosan-based colloidal polyelectrolyte complexes for drug delivery: a review. Carbohyd Polym 238:116126

16. Zhang L et al (2016) Preparation of polyelectrolyte complex nanoparticles of chitosan and poly (2-acry lamido-2-methylpropanesulfonic acid) for doxorubicin release. Mater Sci Eng, C 58:724–729
17. Jha R, Mayanovic RA (2023) A review of the preparation, characterization, and applications of chitosan nanoparticles in nanomedicine. Nanomaterials 13(8):1302
18. Tan C et al (2016) Polysaccharide-based nanoparticles by chitosan and gum arabic polyelectrolyte complexation as carriers for curcumin. Food Hydrocolloids 57:236–245
19. Deng X et al (2021) Stealth PEGylated chitosan polyelectrolyte complex nanoparticles as drug delivery carrier. J Biomater Sci Polym Ed 32(11):1387–1405
20. Jiang L et al (2018) Application of a simple desolvation method to increase the formation yield, physical stability and hydrophobic drug encapsulation capacity of chitosan-based nanoparticles. Int J Pharm 545(1–2):117–127
21. Ammanage AS, Mastiholimath VS (2025) Quality by design approach assisted development and optimization of chitosan-vildagliptin nanoparticles using a simple desolvation technique. J Appl Pharmaceut Sci 15(3):174–182
22. Bashir SM et al (2022) Chitosan nanoparticles: a versatile platform for biomedical applications. Materials 15(19):6521
23. Luque-Alcaraz AG et al (2016) Preparation of chitosan nanoparticles by nanoprecipitation and their ability as a drug nanocarrier. RSC Adv 6(64):59250–59256
24. Loureiro J et al (2022) Single-step self-assembly of zein–honey–chitosan nanoparticles for hydrophilic drug incorporation by flash nanoprecipitation. Pharmaceutics 14(5):920
25. Karimi S, Roshani Farahani M (2024) A single step and cost-effective flash nanoprecipitation method to prepare chitosan with high curcumin conjugated gold nanoparticles loading and its antibacterial investigation. Colloid Nanosci J 2(1):254–260
26. Ngan LTK et al (2014) Preparation of chitosan nanoparticles by spray drying, and their antibacterial activity. Res Chem Intermed 40(6):2165–2175
27. Wang Y et al (2016) Manufacturing techniques and surface engineering of polymer based nanoparticles for targeted drug delivery to cancer. Nanomaterials 6(2):26
28. Nguyen TV et al (2017) Preparation of chitosan nanoparticles by TPP ionic gelation combined with spray drying, and the antibacterial activity of chitosan nanoparticles and a chitosan nanoparticle–amoxicillin complex. Res Chem Intermed 43(6):3527–3537
29. Le Grill S et al (2024) Towards chitosan-amorphous calcium phosphate nanocomposite: co-precipitation induced by spray drying. Next Materials 4:100095
30. Ryan C et al (2017) Synthesis and characterisation of cross-linked chitosan composites functionalised with silver and gold nanoparticles for antimicrobial applications. Sci Technol Adv Mater 18(1):528–540
31. Muñana-González S et al (2023) Covalent cross-linking as a strategy to prepare water-dispersible chitosan nanogels. Polymers 15(2):434
32. Silvestre J et al (2024) Optimization of chitosan adhesive properties by means of genipin crosslinking. J Adhes 100(9):828–844
33. Carmona P et al (2021) Glyceraldehyde as an efficient chemical crosslinker agent for the formation of chitosan hydrogels. Gels 7(4):186
34. Shrestha S, Diogenes A, Kishen A (2014) Temporal-controlled release of bovine serum albumin from chitosan nanoparticles: effect on the regulation of alkaline phosphatase activity in stem cells from apical papilla. J Endodontics 40(9):1349–1354
35. Moghaddam FA, Atyabi F, Dinarvand R (2009) Preparation and in vitro evaluation of mucoadhesion and permeation enhancement of thiolated chitosan-pHEMA core-shell nanoparticles. Nanomed Nanotechnol Biol Med 5(2):208–215
36. Kayitmazer A, Koksal A, Iyilik EK (2015) Complex coacervation of hyaluronic acid and chitosan: effects of pH, ionic strength, charge density, chain length and the charge ratio. Soft Matter 11(44):8605–8612
37. Klein RS et al (2025) Iota-carrageenan/chitosan nanoparticles via coacervation: achieving stability for tiny particles. Nanomaterials 15(3):161

38. Karthikeyan S et al (2017) Proton-conducting I-carrageenan-based biopolymer electrolyte for fuel cell application. Ionics 23(10):2775–2780
39. Garg U et al (2019) Current advances in chitosan nanoparticles based drug delivery and targeting. Adv Pharmaceut Bull 9(2):195
40. Tokumitsu H, Ichikawa H, Fukumori Y (1999) Chitosan-gadopentetic acid complex nanoparticles for gadolinium neutron-capture therapy of cancer: preparation by novel emulsion-droplet coalescence technique and characterization. Pharm Res 16(12):1830–1835
41. Domínguez-Delgado CL et al (2022) Effects of process and formulation parameters on submicron polymeric particles produced by a rapid emulsion-diffusion method. Nanomaterials 12(2):229
42. Pineda-Reyes AM et al (2021) Implementation of the emulsification-diffusion method by solvent displacement for polystyrene nanoparticles prepared from recycled material. RSC Adv 11(4):2226–2234
43. Pulingam T et al (2022) Exploring various techniques for the chemical and biological synthesis of polymeric nanoparticles. Nanomaterials 12(3):576
44. Asal HA et al (2022) Controlled synthesis of in-situ gold nanoparticles onto chitosan functionalized PLGA nanoparticles for oral insulin delivery. Int J Biol Macromol 209:2188–2196

# Chapter 3
# Characterization of Chitosan Nanoparticles

**Abstract** The physicochemical properties of CSNPs are crucial for understanding their stability, structural integrity, and potential applications in biomedicine. The present chapter examines analytical approaches used to quantify CSNPs, focusing on particle dimensions and size distribution through dynamic light scattering (DLS), nanoparticle tracking analysis (NTA), and other relevant microscopic techniques. The AFM enables imaging at atomic resolution, facilitating the evaluation of surface roughness and nanostructures. SEM and TEM are employed for morphological characterization of particles, focusing on particle shape, uniformity, and structure. The discussion also includes zeta potential results to assess surface charge and colloidal stability in physiological medium. The characterization techniques described establish a systematic protocol linking NP characteristics to their functional performance, thereby enhancing reproducibility and enabling the rational design of CS-based nanocarriers.

**Keyword** AFM · Characterization · SEM · TEM · Physicochemical

The physical traits of polymeric nanoparticles (PNPs) vary decisively relying on their composition, quantity, dimensions, surface characteristics, crystalline structure, and dispersed state, among other variables [1]. A comprehensive understanding of the physicochemical, biological, and formation processes of CSNPs is essential for the preparation of NP-based therapeutic formulations and for quality regulation, and characterizing CNPs is crucial. The small size and large surface area render it essential to utilize precise analytical methods in assessing factors including size distribution, morphology, surface charge, chemical composition, and in vitro release profiles [2]. These attributes are essential for evaluating their efficacy in drug delivery systems and are assessed using a variety of characterization methods to achieve a comprehensive evaluation [3]. An in-depth analysis of PNPs requires into consideration both the core particle and the surface ligands, significant in defining the physicochemical aspects and biologic interactions of the molecule. Innovative and advanced methodologies are increasingly utilized to examine the kinetics of NP formation and explore structural asymmetry that profoundly affect their functionality. Notwithstanding considerable

 
A. Haider et al., *Chitosan Nanoparticles for Brucellosis Treatment*, Springer Series in Biomaterials Science and Engineering 21, https://doi.org/10.1007/978-3-032-16292-2_3

**Table 3.1** Parameters along with the appropriate characterization technique

| Parameters | Suitable techniques |
|---|---|
| Crystal structure | HRTEM, XRD, STEM, Electron diffraction |
| NPs dispersion in matrices | AFM, SEM, TEM |
| Agglomeration state | DLS, Zeta Potential, Cryo-EM, DCS |
| Specific surface area | BET |
| Surface charge | EPM, Zeta Potential |
| Particle size distribution | DLS, DTA, SEM, TRPS, DCS |
| Shape | AFM, TEM, HRTEM, Tomography |
| Optical properties size (Structural features) | PL, UV–Vis XRD, SEM, AFRM, TEM, HRTEM, UV–VIS, NMR, DCS |

advancements, numerous obstacles remain in the realm of NP characterization. The challenges encompass issues related to sample preparation, absence of standardized protocols, scarcity of suitable reference components for calibration, and intricacies involved in data interpretation. In the arena of scaled-up production, the necessity for dependable, real time, and in-situ quantifying methods grows ever more critical. Furthermore, characterization of complex matrices continues to represent an unfulfilled requirement [4]. Eventually, the process of characterization holds significant importance in the discipline of nanotoxicology and in evaluating exposure risks involving nanomaterials. Comprehending the behavior of NPs within biological and environmental systems is crucial for assessment of risks and adherence to regulatory standards. Consequently, a comprehensive and multi-faceted characterization approach is essential to address shortcomings of particular methodologies and to assure advancement of safe and effective nano-medicine products [5]. It is possible to characterize CNPs for information relevant to their morphological, optical, and structural features by using a number of complementary approaches (Table 3.1). A diverse array of methodologies is employed to evaluate these parameters, notably Transmission Electron Microscopy (TEM), Scanning Electron Microscopy (SEM), and Atomic Force Microscopy (AFM), which yield comprehensive insights into dimension, shape, and surface appearance of NPs. Dynamic light scattering (DLS), often referred to as Photon Correlation Spectroscopy (PCS), is frequently utilized to ascertain particle size distribution and polydispersity index. A pertinent criterion for evaluating the efficiency of surface changes or drug adsorption procedures is measurement of surface charge, often known as zeta potential, which is used to determine colloidal stability [6].

## 3.1 Size of Chitosan Nanoparticles

CSNP's distinctive physicochemical properties have piqued a considerable intrigue in multiple domains, including drug administration, engineering, genome therapy, and vaccine distribution [7]. Considerable factors affecting biological interactions, including cellular absorption, biological dispersion, excretion, and therapeutic reliance, are governed by particle size. The effectiveness of CSNPs is significantly influenced by their dimensions, which generally range from 10 to 1000 nm, depending on synthesis technique and specified application [8]. The multifarious traits of CSNPs, such as their potential to conjugate with pharmaceuticals, their crystalline composition, and charge, are substantially influenced by their size. The synthesis process, physicochemical features of CS such as molecular mass, deacetylation, and formulation variables such as pH, temperature, and cross-linker all have substantial impact on dimensions of CSNPs, often between 50 and 500 nm [9]. Ionic gelation, being the preferred approach for forming CSNPs owing to its ease, reaction conditions, and lack of organic solvents, usually requires electrostatic interaction between positive amino acid charges of CS and inversely charged polyanions, like tripolyphosphate (TPP). The CS-TPP nanomaterials were first demonstrated via ionic gelation by Calvo et al. [10], who reported particle sizes ranged from 200 to 300 nm, relying on CS/TPP ratio and preparation parameters. It is imperative to contemplate the proportion of CS to TPP, as production of tiny NPs is a direct consequence of increased TPP concentrations, attributed to increased cross-linking density [11]. Meanwhile, MW of CS is crucial as higher MWs usually result in greater size NPs owing to increased viscosity and chain interaction during NP production process. Conversely, CS with lower MW generally yields smaller and more uniform NPs. A meticulous analysis of determinants that influence the dimensions of CS-TPP NPs was conducted in 2018 by Sreekumar. Based on their research, it is feasible to control the size and characteristics of CS particles ranging from nano to microscale by manipulating two essential components of polymer solution: concentration and solvent environment [10]. The size of CSNPs has substantial impact on their biological performance, particularly in terms of therapeutic efficiency, distribution, and cellular absorption. Studies indicate that NPs with diameters under 200 nm are generally considered optimal for systemic delivery, as they can evade rapid clearance by the reticuloendothelial system and achieve improved tissue penetration. In this regard, NPs with dimensions ranging from 100 to 200 nm have shown significant increase in endocytosis by epithelial cells, along with improvement in mucosal uptake [12, 13]. However, NPs larger than 300 nm in size may be advantageous for targeted therapies, such as intratumoral or mucosal drug delivery, where protracted retention at administration site is preferred [14, 15].

Contemporary techniques for synthesizing NPs, as microfluidics and high-pressure homogenization, have allowed for better control of CSNP size and homogeneity [16]. Significant improvements in stability and systemic duration of smaller CSNPs have been achieved through surface modifications, including PEGylation [17]. Enzymatic cross-linking is one novel green synthesis approach that has been studied to produce CSNPs of tunable size with minimal consumption of noxious chemicals [18]. CSNP size is usually determined by using techniques like dynamic light scattering (DLS), scanning electron microscopy (SEM), or transmission electron microscopy (TEM). Significant polydispersity observed in carbon NPs frequently presents challenges when using dynamic light scattering to determine size. Depending on estimated morphology of CSNPs, various imaging-based or non-imaging methods can potentially be adopted to measure particle size. The non-imaging method is grounded in fundamentals of light scattering and comprises both static and dynamic light scattering (DLS). These strategies exhibit wide range of applicability, precision, sensitivity, and efficiency in estimation of particulate sizes [2]. Investigations have confirmed that precisely regulating formulation specifications enables the variation of NP sizes that is indispensable for variety of therapeutic functions. In conclusion, dimensions of CSNPs significantly impact their efficacy and suitability for various biomedical applications. The effectiveness of drug administration, gene therapy, and biological engineering can be enhanced by precisely controlling synthesis variables, particularly within the ionic gelation method, which allows for fine-tuning of particle size to align with specific therapeutic objectives. Scalable production strategies should be the primary focus of future ventures to ensure uniformity and foster clinical applicability. The efficient progression of CS-based nanocarriers that have been tailored for range of biomedical applications is ensured by the comprehension and optimization of these parameters.

## 3.2 Dynamic Light Scattering

A valuable analytical tool for characterizing CSNPs is dynamic light scattering (DLS), also termed as photon correlation spectroscopy (PCS) or quasi-elastic light scattering. To assess the size distribution of suspended particles and their colloidal stability, this approach provides valuable information on hydrodynamic diameter and polydispersity index (PDI) by analyzing intensity changes in light scattered by their Brownian motion. Estimates of hydrodynamic diameters ($D_{\mathrm{h}}$) were made in the Stokes–Einstein equation [19], i.e., $D_{\tau} = k_B T / 6\pi\eta Rh$, in where $T$ stands for absolute temperature and η for viscosity, and kB is the Boltzmann coefficient, which is measured as $1.380 \times 10^{-23}$ kg m$^2$ s$^{-2}$ K$^{-1}$. Implementing two-parameter fitting techniques to correlation data, particularly via cumulant analysis, yields the PDI. This index is dimensionless and calibrated to provide results below 0.05 using monodisperse standards. A sample with PDI values exceeding 0.7 is likely inappropriate for analysis using the DLS technique due to its wide particle size distribution. A variety of size distribution algorithms are implemented with data that falls within 0.05–0.7 values of PDI [20].

Regarding the dimension of particles, polydispersity coefficient, and zeta potential, DLS provides pertinent specifics to gauge the stability as well as reliability of CSNPs for application in biomedical technologies. In order to reduce anomalies as dispersing or consolidation, recent investigations have emphasized the importance of optimizing DLS measurement variables for CSNPs, which include factors such as solvent viscosity, humidity, and concentration [21]. Nano-CS was synthesized by Abdeltwab et al. [22] by cross-linking positively charged CS (M.W = 125.68 kDa) with negatively charged ions TPP using an ionic gelation technique. Based on results of size, distribution, and uniformity analysis conducted using DLS, it was found that majority of nano-CS particles (88%), spanning between 93.76 and 405 nm, were within this range. In distinct analysis, diameter of synthesized low molecular-weight cross-linked CSNPs incorporating 5-FU was assessed through dynamic light scattering, revealing a low PDI value of 0.1 in aqueous phase, which demonstrated significant colloidal stability attributed to the cross-linker glutaraldehyde (GA). Micelle-based NPs, with their size-controllable properties, provide significant advancement in therapeutic efficacy in real time, thereby enhancing bioavailability of encapsulated 5-fluorouracil. The utilization of nanosuspension, along with polydispersity index of less than 0.1, has been shown to enhance the stability of nanocarrier therapeutic agents aimed at liver targeting [23].

A study conducted in 2024 detailed preparation of CS-based semi-interpenetrating network NPs, utilizing (3:1) TPP cross-linker (CS: TPP). Subsequently, free radical polymerization is employed for producing inventive pH-sensitive semi-interpenetrating networks (semi-IPN) consisting of CS-poly (1-vinyl imidazole) (CS-PVIm) and CS-poly(methacrylic acid) (CS-PMAA) NPs [24]. Their DLS assessment reveals that low PDI signifies relatively uniform size distribution. Thus, the CS-TPP NPs exhibit a lesser PDI value of 0.21, originating from 0.25% CS solution, with a mean particle diameter of 63.56 nm (Table 3.2).

The incorporation of MAA in CS-PMAA NPs notably enhances particle size. A comparable pattern was likewise noted for CS-PVIm NPs upon incorporation of VIm. The minimum dimensions recorded for CS-PMAA NPs were 107.9 nm, while CS-PVIm NPs measured 138.32 nm with low PDI values, suggesting a comparatively

**Table 3.2** Formulation, dimensions of particles, and PDI value of synthesized NPs

| Nanoparticles | Compositions (wt/wt/wt) | | | | | Particle size (nm) |
|---|---|---|---|---|---|---|
| | CS | TPP | PMAA | PVIm | MBA | |
| CS-TPP NPs(0.25% of CS sol) | 3 | 1 | – | – | – | 58.3 |
| CS-TPP (0.125% of CS sol) | 3 | 1 | – | – | – | 63.56 |
| CS-TPP (0.50% of CS sol) | 3 | 1 | – | – | 15 | 99.3 |
| CS-PMAA | 100 | – | 53 | – | 15 | 107.9 |
| CS-PMAA | 100 | – | 105 | – | 15 | 102.0 |
| CS-PMAA | 100 | – | 210 | – | 15 | 232.9 |
| CS-PVIm | 100 | – | – | 53 | 15 | 138.32 |
| CS-PVIm | 100 | – | – | 117 | 15 | 179.7 |
| CS-PVIm | 100 | – | – | 175 | 15 | 211.2 |

Reproduced with permission from Ref. [24]

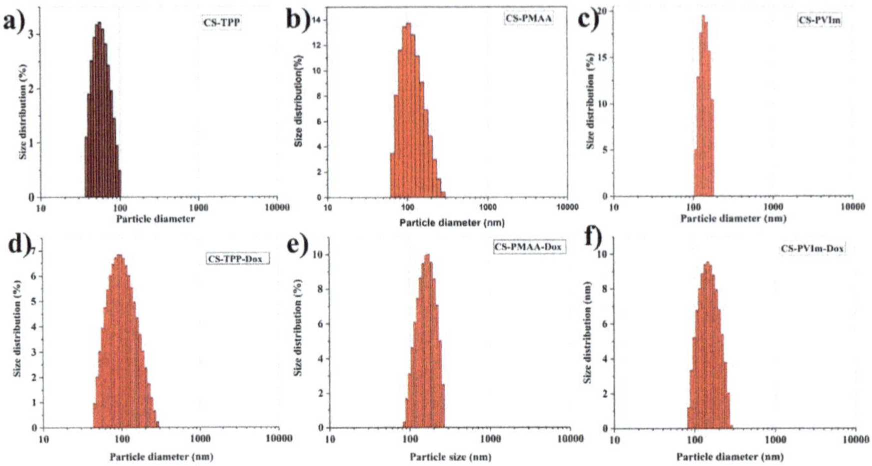


**Fig. 3.1** Hydrodynamic particle size of **a** CS-TPP, **b** CS-PMAA, **c** CS-PVIm, **d** CS-TPP-Dox, **e** CS-PMAA-Dox, and **f** CS-PVIm-Dox nanoparticles. Reproduced with permission from Ref. [24]

uniform size distribution. The mean hydrodynamic particle sizes of Dox-loaded CS-TPP, CS-PMAA, and CS-PVIm NPs were measured at 100.2, 175.2, and 165.84 nm, as depicted in Fig. 3.1.

Gamboa et al. [25] produced CSNPs through ionotropic gelation method, utilizing TPP as cross-linking agent in various ratios ranging from 6:1 to 3:1. The average hydrodynamic diameters exhibited range of 178–205 nm, demonstrating variability in accordance with CS: TPP ratio. The NPs analyzed through dynamic light scattering reveal that increased concentration of TPP correlates with reduction in particle size, evidenced by average hydrodynamic diameter.

## 3.3 Atomic Force Microscopy

The comprehensive examination of nano-sized geometry and surface layout of CSNPs is facilitated by employing Atomic Force Microscopy (AFM). This methodology sustains the intrinsic state of sample and enables high-resolution imaging, thereby eradicating the need for conductive coatings or vacuum surroundings. The functionality and efficiency of CSNPs for a variety of biological fields are significantly influenced by morphology, particle distribution, and surface texture of NPs, which can be examined thoroughly via AFM. Recent investigations have substantiated the effectiveness of AFM for thorough characterization of nanostructures in verification of morphology, dimensions (size), assessment of surface roughness, and their effectiveness in liposomal drug delivery applications [26]. In 2021, Webber et al. used κ- and λ-carrageenan with higher MW to formulate carrageenan/chitosan multilayers with nisin Z embedded. Peak force tapping mode was adopted to determine topography and sample thickness. Overall, the samples had comparable heights, while the sample without nisin Z exhibited the most roughness [27].

Bhirud and his co-authors in 2025 utilized AFM (5300E, Hitachi, Japan) in semi-contact mode, with high vacuum to evaluate surface morphology and topographical aspects of chitosan-based NPs with coated Tocopheryl Polyethylene Glycol 1000 Succinate for imatinib mesylate targeted delivery (CS-IMTTPGS). Analysis included assessment of 3-D scans, average roughness, height, and asymmetrical patterns [28]. The AFM analyzes particle size of 198 nm, with particle size distribution demonstrating skewness of $0.596 \pm 0.22$, suggesting a slight tendency for larger particle sizes (Fig. 3.2). Nonetheless, the comparatively modest skewness value suggests a distribution that is quite symmetrical. Surface roughness is key aspect in interaction between NPs and biosystems. The root mean square (RMS) value was employed to quantify this attribute, and particles exhibit a relatively irregular topography, with an average roughness of $78.5 \pm 3.11$ nm. A kurtosis value of $0.875 \pm 1.05$ was noticed on average.

Subsequent research has elucidated application of AFM in examination of CSNPs integrated with functional additives. Kızılkaya and Kaya [29] synthesized CS/$TiO_2$/rosmarinic acid NPs. Their AFM examination revealed that incorporating $TiO_2$ and rosmarinic acid into CS matrix resulted in more uniform and refined surface morphology, exhibiting topographic height variations ranging from 0 to 20.4 nm.

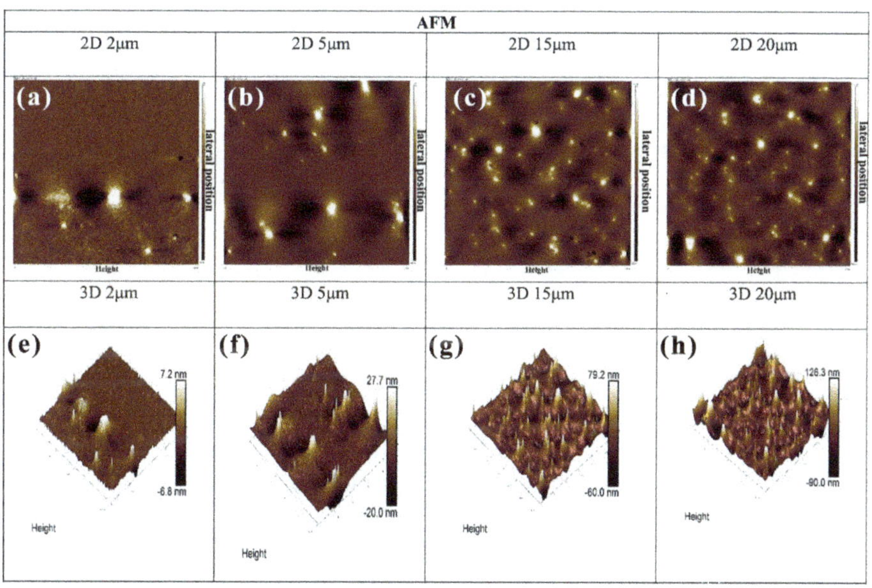


**Fig. 3.2** AFM images of CS-IMT-TPGS NPs: **a–d** 2-D, and **e–h** 3-D. Reproduced with authorization from Ref. [28]

Homogeneity of surface is essential in fields such as antimicrobial coatings, alongside characteristics of surface can significantly affect bacterial adhesion and formation of biofilms. Furthermore, employing AFM with scan size of 50 × 50 µm at a resolution of less than 0.1 nm, Hosseini et al. [30] assessed morphology of bioactive Fish Gelatin/CSNPs films and influence of oregano essential oil, while also determining their average roughness ($R_a$) and root mean-square roughness ($R_q$). AFM images of CSNPs reveal consistently sized 40–80 nm spherical particles. The oil-free film exhibits a surface that is comparatively coarse, characterized by 94.56 and 118.90 nm values of Ra and Rq. The inclusion of essential oil resulted in notable reductions ($p < 0.05$) in the roughness of the films, as evidenced by diminished Ra and Rq values (25.83 and 33.63 nm, respectively, for 0.4% (w/v) essential oil) as illustrated in Fig. 3.3.

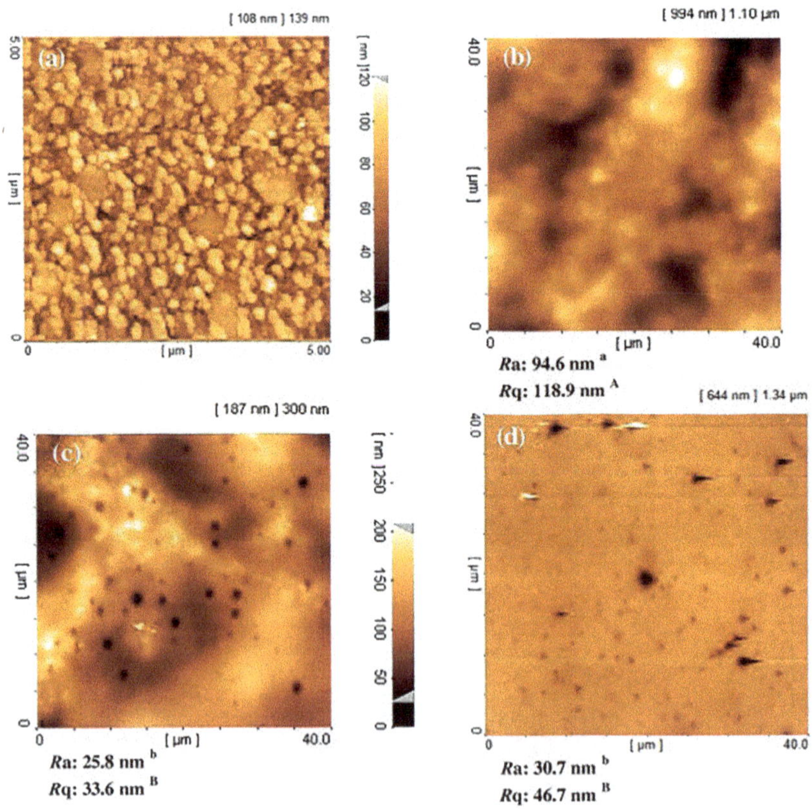


**Fig. 3.3** AFM scan of **a** CSNP, **b** control and FG/CSNPs composite films containing, **c** 0.4, and **d** 1.2% (w/v) OEO (black dots in image). Claimed with authorization from Ref. [30]

These instances in tandem demonstrate adaptability of AFM in assessing essential physical attributes of CSNPs, including dimensions, surface texture, and uniformity of coatings. Through the provision of intricate nanoscale surface information, AFM facilitates refinement of synthesis techniques and aids in establishing connections between physicochemical characteristics and biological efficacy, thus informing methodical design of CS-based nanocarriers for precise and effective therapeutic applications. As high-speed AFM and hybrid methodologies such as AFM-IR continue to evolve, forthcoming research has potential to deepen our understanding of structure–function dynamics of CSNPs, particularly in realms of biomedical and commercial uses.

## 3.4 Nanoparticle Tracking Analysis

NP Tracking Analysis (NPTA) has become an essential method for the characterization of CSNPs, providing immediate representation and quantifying particle size distribution and concentration within colloidal suspensions. In contrast to conventional techniques like dynamic light scattering (DLS), NPTA monitors individual particles, offering nuanced understanding of polydispersity and aggregation dynamics. Recent studies have elucidated the efficacy of NPTA in enhancing CS NP formulations for applications in drug delivery, gene therapies, and antimicrobial strategies. Chuah et al. [31] employed NPTA to test curcumin-loaded CSNPs for mucoadhesion and drug release in simulated colonic environment. According to the study, nanoparticles exposed to mucin tended to spontaneously cluster, leading to 81% curcumin release in 6 h. within 7.4 pH phosphate buffer. NPTA's ability to analyze NP dynamics in physiologically relevant environments can be verified by aggregation, owing to electrostatic interactions between positive charge on CS and negative charge mucin.

In a comparable manner, Mazzarino et al. adopted NPTA to investigate polycaprolactone NPs that were capped with mucoadhesive polysaccharide CS. These NPs contained curcumin, and their primary purpose was to facilitate buccal distribution of this pharmacological drug (Fig. 3.4). NPs showed homogeneous size distributions, polydispersity indices < 0.3, and encapsulation effectiveness above 99%, according to NPTA findings that were in agreement with DLS data. The importance of NPTA in improving formulations for mucosal drug delivery is highlighted by its validation of stability and homogeneity of NPs [32].

Moreover, in the advancement of CS-based nano-drug delivery systems for interferon alpha, NPTA is utilized to measure particulate concentration and observe Brownian motion. The methodology yielded accurate assessments of hydrodynamic diameter and zeta potential, which are critical factors for ensuring the effectiveness and endurance of NPs in therapeutic contexts [33]. NPTA provides very precise and adaptive analysis of size, concentration, and aggregation behavior, making it a crucial tool for developing and evaluating CS-based nanotechnology for targeted drug delivery.

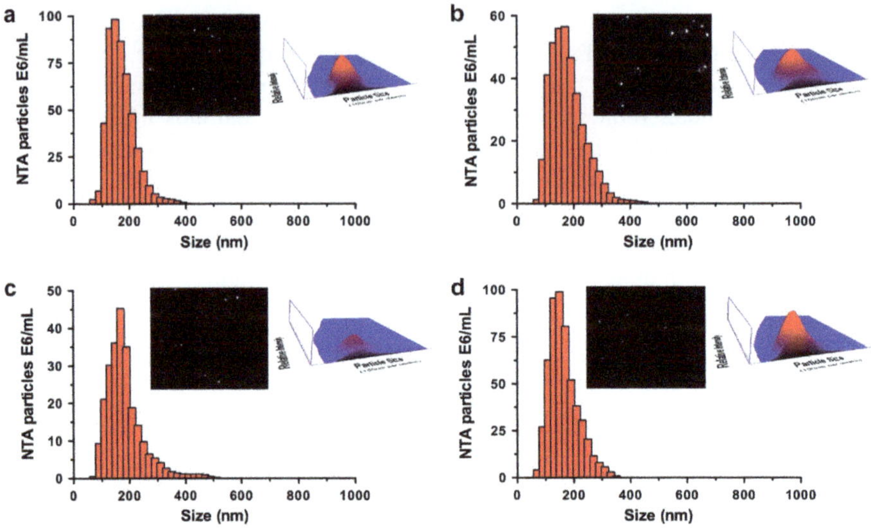


**Fig. 3.4** Size distribution from NTA analysis: Cur (**a**), Cur-CSL (**b**), Cur-CSM (**c**), and Cur-CSH (**d**). Used with permission from Ref. [32]

## 3.5 Surface Morphology of CNSPs

The surface morphology of CSNPs contributes a pivotal role in determining their physical, chemical stability, drug encapsulation effectiveness, and interactions with biological structures. Typically, CSNPs display spherical or roughly spherical morphology, characterized by either smooth or marginally textured surfaces, contingent upon synthesis techniques and formulation parameters employed. In this way, CSNPs synthesized through ionic gelation utilizing sodium TPP as cross-linking agent generally yield uniform and spherical particles, attributed to regulated ionic interactions among chitosan's positively charged amino acids and negative phosphate groups of TPP [34]. The MW and degree of deacetylation of CS can profoundly influence surface characteristics of NPs. Specifically, CS with smaller molecular mass generally results in more uniform and smoother particles, whereas CS with higher molecular weight is associated with irregular or accumulated morphologies. Furthermore, surface morphology may be influenced by integration of various agents or coating components. Folate-conjugated CSNPs demonstrated subtly altered surface design, enhancing nanoparticles' capacity to engage with specific cells within drug delivery framework [35]. To improve efficacy of clinical applications of composite films, it is essential to formulate materials exhibiting diverse physical and chemical properties, as well as surface morphologies that promote adhesion and cellular proliferation. The alteration of CS through incorporation of nano-sized fillers with diverse characteristics, forms, and dimensions, along with synthesis of composites, represents a highly effective approach to modifying properties of CS [36]. The findings indicate that meticulous refinement of formulation elements and processing parameters is crucial for attaining intended surface properties of CSNPs, especially in context of biomedical and pharmaceutical uses. Grasping these morphological features is crucial for improving nanoparticle efficacy in *vivo.*

### *3.5.1 Scanning Electron Microscopy*

The surface appearance of CSNPs can be well described leveraging scanning electron microscopy, which provides substantial details about their size distribution, shape, and morphology at the nano-level. The latest developments in scanning electron microscopy, particularly field-emission SEM and environmental SEM, have empowered researchers to capture high-resolution images of CSNP with reduced sample loss. A thorough investigation revealed that CS-TPP NPs, synthesized through ionic gelation method, exhibited spherical morphology and measured approximately 200 nm in size, devoid of agglomerations. This was confirmed via scanning electron microscopy (CROSS-BEAM 550, ZEISS) at acceleration voltage of 5 kV, utilizing scales of 5.06 and 14.75 Kx [37]. The alteration of CS's molecular mass had pronounced effect on SEM morphology; NPs obtained from CS of lower MW exhibited smaller and more uniform traits, in contrast to those derived from higher MW variants, which were prone to forming larger and more inconsistent aggregates. Barradas et al. [38] employed coacervation method for assembling NPs, noting that resultant size and surface charge are influenced by molecular weight of CS and nitrogen/phosphorus (N/P) ratio. The morphology of NPs, as analyzed by scanning electron microscopy at accelerating voltage of 20 keV, is depicted in Fig. 3.5. The image reveals aggregated spherical NPs, with individual sizes ranging from 179 to 264 nm for the sample labeled 57.5NPs. The 20.6NPs exhibit spherical aggregation pattern of NPs, with sizes ranging from 73.2 to 122 nm.

Research conducted by Areej et al. [39] employs chemical co-precipitation method to synthesize stannous selenide (SnSe) in conjunction with eudragit L100 (Eud) and varying concentrations of CS. The FE-SEM images revealed external topography of both pure and doped nanostructures, showcasing orthorhombic and hexagonal modes characterized by heterogeneous areas for SnSe. The 2% CS/Eud-SnSe exhibited diminishment in size accompanied by augmentation in surface area. The additional introduction of CS resulted in agglomeration and an increase in density, suggesting capping effect of CS. A different investigation synthesized heterojunction of cobalt oxide and zinc oxide ($Co_3O_4$/ZnO) utilizing varying concentrations (2 and 4 wt%) of CS, while maintaining constant amount (3 wt%) of polyvinylpyrrolidone (PVP) through co-precipitation technique. The analysis conducted using FE-SEM revealed presence of irregularly shaped nanoparticles of $Co_3O_4$/ZnO nanocomposites on surface. The inclusion of CS resulted in detection of aggregated flakes and chunks that erupt from surface [40]. SEM continues to be an essential instrument for elucidating physical integrity and surface architecture of CSNPs, thereby aiding in refinement of NP formulations for applications in drug delivery, antimicrobial strategies, and tissue engineering.

**Fig. 3.5** Morphology of NPs determined by SEM. 57.5NPs at 15,000 x (**a**) and 45,000 x (**b**); 20.6NPs at 15,000 x (**c**) and 35,000 x (**d**), separately. Reproduced with permission from Ref. [38]

### *3.5.2 Transmission Electron Microscopy*

Transmission electron microscopy (TEM) has emerged as essential analytical method for elucidating inner structure, dimensions, morphology, and distribution of CSNPs with resolution at the nanometer level that exceeds that of alternative imaging modalities. In TEM, a stream of high-energy electrons interacts with specimens, producing an image that is observable on a fluorescent display. Upon passage of the irradiated electron beam within the sample, both transmitted and diffracted beams are produced, resulting in formation of an image through interference. The image may be concentrated on an imaging apparatus, including a fluorescent screen or charge-coupled device (CCD) camera. This enables high magnification of around 2 Å, and the sample ought to be of slender nature, probably 500 Å in thickness. The preparation of materials for TEM analysis may be achieved through deposition of dilute suspensions onto carbon-coated copper grids, or alternatively, by employing a negative staining agent such as uranyl acetate [41]. In contrast to surface-based methods, transmission microscopy provides high-resolution imaging that facilitates examination of particle uniformity, core–shell configurations, and internal density fluctuations. CSNPs generally display spherical or nearly spherical shapes when observed under TEM, with size distributions frequently varying from 50 to 200 nm, contingent upon formulation parameters employed. In study [5], TEM was employed to examine the morphology and dimensions of the magnetic CSNPs. The $Fe_3O_4$ particles employed in this investigation were effectively coated with chitosan, exhibiting a diameter of CSNPs that varied between 10 and 80 nm. The transmission electron microscopy images of green synthesized carbon NPs, derived from the aqueous extract of fresh *Eucalyptus globulus Labill* leaves, revealed a spherical morphology with dimensions ranging from 6.92 to 10.10 nm [42].

In accordance with requirements of targeted drug delivery, NPs of reduced size demonstrate superior penetration compared to their larger counterparts and possess the ability to traverse capillaries and tissue sinusoids. A research initiative has established a drug delivery mechanism employing folic acid-functionalized CS alongside poly(lactic-co-glycolic acid) nanocarriers, aimed at precise delivery of sodium butyrate to leukemia cells [43]. Their TEM analysis revealed a core size spanning 40–60 nm, exhibiting a spherical shape and homogeneous distribution. This characteristic is beneficial for drug delivery, as sphere-shaped and nanoscale drug delivery systems are more adept at facilitating efficient permeation into cancerous tissues.

Moreover, chemical modification by altering the functional group of nanoparticle surface also influences dimensions of NPs. Summan et al. [44] investigate integration of diverse concentrations of CS and cetyltrimethylammonium bromide (CTAB) within tin selenide (SnSe) to establish an innovative ternary system of CS/CTAB-SnSe through co-precipitation method. TEM images illustrate that introduction of CS into binary system clarifies capping of NSs, with CS serving as capping agent that diminishes size of NRs and NPs. The elevated concentrations of CS resulted in formation of agglomerates and network of interconnected NSs, thereby facilitating swift carrier transport during catalytic reactions.

In separate investigations, manganese dioxide ($MnO_2$) nanorods and CS-grafted polyacrylic acid (CS-g-PAA) doped $MnO_2$ were synthesized through hydrothermal methods. The TEM analysis revealed that $MnO_2$ exhibits nanorod formations, while CS-g-PAA doped $MnO_2$ demonstrates structures resembling nanoflakes (Fig. 3.6). The clustering of nanoflakes diminished with additional rise in dopant concentration, resulting in formation of stacked flower-like morphology. Figure 3.6f illustrates that pronounced agglomeration of nanoflake-like structure offers an extensive surface area, thereby enhancing flow of charge carriers during catalysis [45]. The results elucidate efficacy of TEM in validating nanoscale dimensions, ensuring internal consistency, and assessing potential aggregation behavior, all of which are essential for advancement of CNPs aimed at drug delivery, imaging, and biomedical uses.

## 3.6 Surface Charge of Chitosan Nanoparticles

The durability of CSNPs, their capacity to be taken up by cells, their mucoadhesiveness, and way of interaction with biological membranes are all significantly influenced by their surface charge. The measurement of zeta potential reveals that

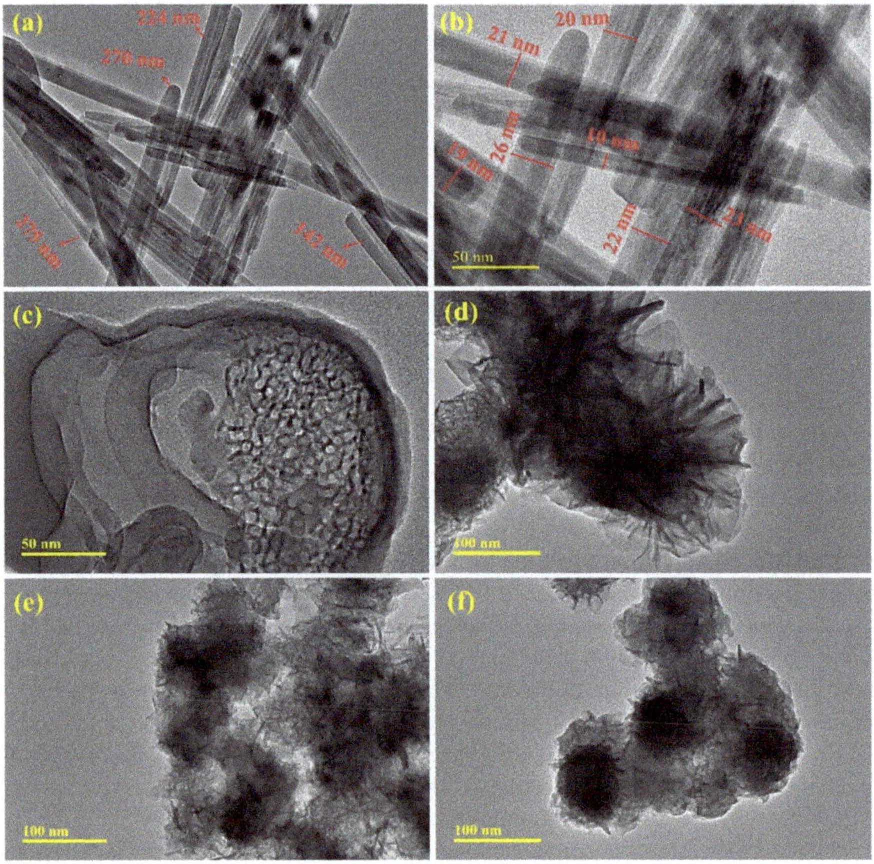


**Fig. 3.6** **a**, **b** TEM micrographs of $MnO_2$, **c** CS-g-PAA, and **d–f** CS-g-PAA (3, 6, and 9 mL) doped $MnO_2$ NSs. Reproduced with permission from Ref. [45]

CS, characterized as a cationic polymer owing to its amine groups, typically confers positive surface charge upon NPs when prepared in acidic or mildly neutral environments. The presence of this positive charge amplifies electrostatic interactions with negatively charged cellular membranes and mucosal surfaces, thereby enhancing bio-adhesion and facilitating intracellular delivery. Current studies indicate that unaltered CSNPs typically display zeta potentials between +20 and +60 mV, contingent upon the extent of deacetylation and the molecular size of CS employed. A recent investigation indicates that the formation of a CS-based NP system, in which CS concentration plays a significant role in enhancing the surface charge of NPs that was ascertained through the degree of interaction with the cationic CS group ($–NH_3^+$) and polyanion group. The elevation in CS concentration resulted in an augmented zeta potential. The concentration of extract solvent notably influenced the zeta potential of NPs, leading to a reduction in zeta potential as the concentration of extract solvent increased. The presence of cation groups diminished due to interactions with polyphenolic compounds in extract, which led to a reduction in surface charge of NPs [46].

Beyond that, pH serves as a principal determinant influencing both particle size and zeta potential of CSNPs. The particle size exhibited a rapid increase from pH 1 to 3.5, followed by a gradual decrease from pH 3.5 to 5.5. The influence of pH on zeta potential exhibited a comparable pattern, characterized by a modest increase at pH 4. The zeta potential value rose from pH 1 to 4, followed by a gradual decline from pH 4 to 5.5. The protonation of amine groups of CS under acidic conditions resulted in significant charge repulsion and molecular expansion [2]. To improve the efficiency of CS-based NPs for anticancer drug delivery that responds to stimuli, Shikuku et al. [24] used free radical polymerization to develop novel pH-sensitive semi-interpenetrating networks (semi-IPN) comprised of CS-poly(1-vinyl imidazole) (CS-PVIm) and CS-poly(methacrylic acid) (CS-PMAA) NPs. The zeta potential of CS-TPP NPs was determined to be + 28.5 mV, as illustrated in Fig. 3.7a. This phenomenon can be ascribed to the cationic characteristics of CS, particularly in conditions of low pH. The zeta potentials for CS-PMAA (Fig. 3.7b) and CS-PVIm NPs (Fig. 3.7c) were observed to range from + 17.2 mV to + 38.3 mV, respectively. When the pH is below 6.5, NPs composed of 0.25% CS and tripolyphosphate exhibit positive zeta potential. This occurs due to the protonation of $-NH_2$ groups of ionized CS, transforming them into $NH_3^+$ under these conditions, which results in elevated zeta potential. Conversely, CS-PMAA exhibits a comparatively diminished zeta potential, attributable to the presence of the $-COO^-$ group, which effectively reduces the surface positive charge of CS. Furthermore, their surface can be charged either positively or negatively by regulating pH.

In a similar vein, Athavale et al. [47] investigated pH-dependent structural characteristics and surface charge density actions of CSNPs. The findings demonstrated that CSNPs exhibit considerable stability at acidic pH levels, whereas they experience a decline in stability under neutral and basic pH conditions, leading to amalgamation. Ramírez et al. [48] posit that alterations in surface characteristics of CSNPs may influence their diffusion within dermal tissue. Thereby, CSNPs were synthesized, and their surface underwent modification with polyethylene glycol, oxalic acid, and

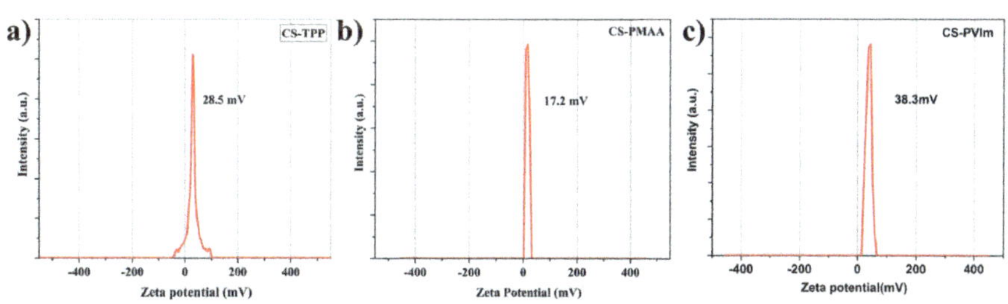


**Fig. 3.7** Zeta potential of **a** CS-TPP, **b** CS-PVIm, and **c** CS-PMAA semi-IPN nanoparticles. Reproduced with permission from Ref. [24]

linoleic acid to facilitate transdermal therapy. Surface modification reveals that the presence of positive (zeta potential + 27.5 mV) or negative (zeta potential − 2.2 mV) surface charge does not influence diffusion of CSNPs. The optimization of CNPs for tailored biomedical applications, as targeted drug delivery, vaccine transporters, and gene delivery systems, is contingent upon adaptability of surface charge, which is vital for charge-dependent interactions with biological limitations.

## 3.7 Conclusion

Nanotechnology persistently transforms numerous fields, encompassing environmental cleanup, biomedicine, and energy, with polymeric NPs, particularly CS-based systems, leading this groundbreaking advancement. CSNPs, due to their compatibility with biological systems, ability to decompose naturally, and diverse functional properties, have attracted considerable interest for use in delivering drugs, therapies, and environmental uses. An essential component of their advancement resides in their comprehensive physicochemical characterization. Grasping the nuances of parameters like surface charge, particle size, morphology, and structural characteristics is essential for both meticulous synthesis and effective application. The various techniques, each possessing distinct advantages and drawbacks, highlight the necessity of a cohesive characterization approach to attain comprehensive understanding of CSNPs. As nanotechnology progresses, its future efficacy will hinge not solely on pioneering material design but also on refinement and enhancement of accurate, multi-faceted characterization methodologies.

Notwithstanding the progress made, obstacles persist in attaining consistent NP synthesis and dependable characterization. Future investigations ought to prioritize establishment of standardized protocols and integration of complementary methodologies to improve both reproducibility and effectiveness of applications. The ongoing advancements in nanotechnology rely on meticulous material characterization, facilitating development of optimized CSNPs for applications in biomedical and environmental domains. This methodical approach guarantees that CSNPs can be customized for particular applications, thereby enhancing their significance in advanced technological developments.

## References

1. Zielińska A, Carreiró F, Oliveira AM, Neves A, Pires B, Venkatesh DN, Durazzo A, Lucarini M, Eder P, Silva AM, Santini A, Souto EB (2020) Polymeric nanoparticles: production, characterization, toxicology and ecotoxicology. Molecules 25:3731. https://doi.org/10.3390/molecules25163731
2. Warsito MF, Agustiani F (2021) A review on factors affecting chitosan nanoparticles formation. IOP Conf Ser Mater Sci Eng 1011:012027. https://doi.org/10.1088/1757-899X/1011/1/012027

3. Eltaib L (2025) Polymeric nanoparticles in targeted drug delivery: unveiling the impact of polymer characterization and fabrication. Polymers 17:833. https://doi.org/10.3390/polym17070833
4. Mourdikoudis S, Pallares RM, Thanh NTK (2018) Characterization techniques for nanoparticles: comparison and complementarity upon studying nanoparticle properties. Nanoscale 10:12871–12934. https://doi.org/10.1039/C8NR02278J
5. Jha R, Mayanovic RA (2023) A review of the preparation, characterization, and applications of chitosan nanoparticles in nanomedicine. Nanomaterials 13:1302. https://doi.org/10.3390/nano13081302
6. Guterres SS, Alves MP, Pohlmann AR (2007) Polymeric nanoparticles, nanospheres and nanocapsules, for cutaneous applications. Drug Target Insights 2:117739280700200. https://doi.org/10.1177/117739280700200002
7. Mohammed M, Syeda J, Wasan K, Wasan E (2017) An overview of chitosan nanoparticles and its application in non-parenteral drug delivery. Pharmaceutics 9:53. https://doi.org/10.3390/pharmaceutics9040053
8. Agnihotri SA, Mallikarjuna NN, Aminabhavi TM (2004) Recent advances on chitosan-based micro- and nanoparticles in drug delivery. J Control Release 100:5–28. https://doi.org/10.1016/j.jconrel.2004.08.010
9. Fan W, Yan W, Xu Z, Ni H (2012) Formation mechanism of monodisperse, low molecular weight chitosan nanoparticles by ionic gelation technique. Colloids Surf, B 90:21–27. https://doi.org/10.1016/j.colsurfb.2011.09.042
10. Sreekumar S, Goycoolea FM, Moerschbacher BM, Rivera-Rodriguez GR (2018) Parameters influencing the size of chitosan-TPP nano- and microparticles. Sci Rep 8:4695. https://doi.org/10.1038/s41598-018-23064-4
11. Gan Q, Wang T, Cochrane C, McCarron P (2005) Modulation of surface charge, particle size and morphological properties of chitosan–TPP nanoparticles intended for gene delivery. Colloids Surf, B 44:65–73. https://doi.org/10.1016/j.colsurfb.2005.06.001
12. Alexis F, Pridgen E, Molnar LK, Farokhzad OC (2008) Factors affecting the clearance and biodistribution of polymeric nanoparticles. Mol Pharm 5:505–515. https://doi.org/10.1021/mp800051m
13. Hoshyar N, Gray S, Han H, Bao G (2016) The effect of nanoparticle size on in vivo pharmacokinetics and cellular interaction. Nanomedicine 11:673–692. https://doi.org/10.2217/nnm.16.5
14. Kumari A, Yadav SK, Yadav SC (2010) Biodegradable polymeric nanoparticles based drug delivery systems. Colloids Surf, B 75:1–18. https://doi.org/10.1016/j.colsurfb.2009.09.001
15. Islam S, Ahmed MMS, Islam MA, Hossain N, Chowdhury MA (2025) Advances in nanoparticles in targeted drug delivery–a review. Results Surf Interfaces 19:100529. https://doi.org/10.1016/j.rsurfi.2025.100529
16. Nunziata G, Borroni A, Rossi F (2025) Advanced microfluidic strategies for core-shell nanoparticles: the next-generation of polymeric and lipid-based drug nanocarriers. Chem Eng J Adv 22:100759. https://doi.org/10.1016/j.ceja.2025.100759
17. Suk JS, Xu Q, Kim N, Hanes J, Ensign LM (2016) PEGylation as a strategy for improving nanoparticle-based drug and gene delivery. Adv Drug Deliv Rev 99:28–51. https://doi.org/10.1016/j.addr.2015.09.012
18. Abuzeid HM, Julien CM, Zhu L, Hashem AM (2023) Green synthesis of nanoparticles and their energy storage, environmental, and biomedical applications. Crystals 13:1576. https://doi.org/10.3390/cryst13111576
19. Stetefeld J, McKenna SA, Patel TR (2016) Dynamic light scattering: a practical guide and applications in biomedical sciences. Biophys Rev 8:409–427. https://doi.org/10.1007/s12551-016-0218-6
20. Danaei M, Dehghankhold M, Ataei S, Hasanzadeh Davarani F, Javanmard R, Dokhani A, Khorasani S, Mozafari MR (2018) Impact of particle size and polydispersity index on the clinical applications of lipidic nanocarrier systems. Pharmaceutics 10:57. https://doi.org/10.3390/pharmaceutics10020057

21. Rodriguez-Loya J, Lerma M, Gardea-Torresdey JL (2023) Dynamic light scattering and its application to control nanoparticle aggregation in colloidal systems: a review. Micromachines 15:24. https://doi.org/10.3390/mi15010024
22. Abdeltwab WM et al (2019) Antimicrobial effect of chitosan and nano-chitosan against some pathogens and spoilage microorganisms. J Adv Lab Res Biol 10(1):8–15
23. Sethi A, Ahmad M, Huma T, Khalid I, Ahmad I (2021) Evaluation of low molecular weight cross linked chitosan nanoparticles, to enhance the bioavailability of 5-flourouracil. Dose-Response 19. https://doi.org/10.1177/15593258211025353
24. Shikuku R, Hasnat MA, Mashrur SBA, Haque P, Rahman MM, Khan MN (2024) Chitosan-based pH-sensitive semi-interpenetrating network nanoparticles as a sustained release matrix for anticancer drug delivery. Carbohydr Polym Technol Appl 7:100515. https://doi.org/10.1016/j.carpta.2024.100515
25. Des Bouillons-Gamboa RE, Montes de Oca G, Baudrit JRV, Ríos Duarte LC, Lopretti M, Rentería Urquiza M, Zúñiga-Umaña JM, Barreiro F, Vázquez P (2024) Synthesis of chitosan nanoparticles (CSNP): effect of CH-CH-TPP ratio on size and stability of NPs. Front Chem 12:1469271. https://doi.org/10.3389/fchem.2024.1469271
26. Joshi J, Homburg SV, Ehrmann A (2022) Atomic Force Microscopy (AFM) on biopolymers and hydrogels for biotechnological applications—possibilities and limits. Polymers 14:1267. https://doi.org/10.3390/polym14061267
27. Webber JL, Namivandi-Zangeneh R, Drozdek S, Wilk KA, Boyer C, Wong EHH, Bradshaw-Hajek BH, Krasowska M, Beattie DA (2021) Incorporation and antimicrobial activity of nisin Z within carrageenan/chitosan multilayers. Sci Rep 11:1690. https://doi.org/10.1038/s41598-020-79702-3
28. Bhirud D, Bhattacharya S, Raval H, Sangave PC, Gupta GL, Paraskar G, Jha M, Sharma S, Belemkar S, Kumar D, Maheshwari R, Sharma M (2025) Chitosan nanoparticles of imatinib mesylate coated with TPGS for the treatment of colon cancer: in-vivo & in-vitro studies. Carbohyd Polym 348:122935. https://doi.org/10.1016/j.carbpol.2024.122935
29. Kızılkaya P, Kaya M (2024) Chitosan/$TiO_2$/rosmarinic acid bio-nanocomposite coatings: characterization and preparation. J Compos Sci 9:2. https://doi.org/10.3390/jcs9010002
30. Hosseini SF, Rezaei M, Zandi M, Farahmandghavi F (2016) Development of bioactive fish gelatin/chitosan nanoparticles composite films with antimicrobial properties. Food Chem 194:1266–1274. https://doi.org/10.1016/j.foodchem.2015.09.004
31. Chuah LH, Roberts CJ, Billa N, Abdullah S, Rosli R, Manickam S (2014) Using Nanoparticle Tracking Analysis (NTA) to decipher mucoadhesion propensity of curcumin-containing chitosan nanoparticles and curcumin release. J Dispersion Sci Technol 35:1201–1207. https://doi.org/10.1080/01932691.2013.800458
32. Mazzarino L, Travelet C, Ortega-Murillo S, Otsuka I, Pignot-Paintrand I, Lemos-Senna E, Borsali R (2012) Elaboration of chitosan-coated nanoparticles loaded with curcumin for mucoadhesive applications. J Colloid Interface Sci 370:58–66. https://doi.org/10.1016/j.jcis.2011.12.063
33. Imperiale JC, Schlachet I, Lewicki M, Sosnik A, Biglione MM (2019) Oral pharmacokinetics of a chitosan-based nano-drug delivery system of interferon alpha. Polymers 11:1862. https://doi.org/10.3390/polym11111862
34. Hu Q, Luo Y (2021) Chitosan-based nanocarriers for encapsulation and delivery of curcumin: a review. Int J Biol Macromol 179:125–135. https://doi.org/10.1016/j.ijbiomac.2021.02.216
35. Kesharwani P, Halwai K, Jha SK, Mughram MHAL, Almujri SS, Almalki WH, Sahebkar A (2024) Folate-engineered chitosan nanoparticles: next-generation anticancer nanocarriers. Mol Cancer 23:244. https://doi.org/10.1186/s12943-024-02163-z
36. Dresvyanina EN, Kodolova-Chukhontseva VV, Bystrov SG, Dobrovolskaya IP, Vaganov GV, Smirnova NV, Kolbe KA, Kamalov AM, Ivan'kova EM, Morganti P, Yudin VE (2021) Influence of surface morphology of chitosan films modified by chitin nanofibrils on their biological properties. Carbohydr Polym 262:117917. https://doi.org/10.1016/j.carbpol.2021.117917
37. Gutiérrez-Ruíz SC, Cortes H, González-Torres M, Almarhoon ZM, Gürer ES, Sharifi-Rad J, Leyva-Gómez G (2024) Optimize the parameters for the synthesis by the ionic gelation

technique, purification, and freeze-drying of chitosan-sodium tripolyphosphate nanoparticles for biomedical purposes. J Biol Eng 18:12. https://doi.org/10.1186/s13036-024-00403-w
38. Aranda-Barradas ME, Trejo-López SE, Del Real A, Álvarez-Almazán S, Méndez-Albores A, García-Tovar CG, González-Díaz FR, Miranda-Castro SP (2022) Effect of molecular weight of chitosan on the physicochemical, morphological, and biological properties of polyplex nanoparticles intended for gene delivery. Carbohydr Polym Technol Appl 4:100228. https://doi.org/10.1016/j.carpta.2022.100228
39. Fatima A, Naz M, Haider A, Shahzadi I, Ul-Hamid A, Ullah H, Ali G, Moeen S, El-Bahy SM, Ikram M (2025) DFT, molecular docking, antimicrobial activity, dye degradation, and oxygen evolution reaction of functionalized chitosan integrated eudragit-stannous selenide nanostructures. Surf Interfaces 58:105907. https://doi.org/10.1016/j.surfin.2025.105907
40. Ullah H, Dilpazir S, Haider A, Shahzadi A, Ul-Hamid A, Ali G, Hussain S, Fouda AM, Ikram M (2025) Chitosan and PVP doped cobalt/zinc catalysts for dye degradation and antimicrobial activity supported by molecular docking analysis. Mater Today Commun 42:111150. https://doi.org/10.1016/j.mtcomm.2024.111150
41. Chandra Hembram K, Prabha S, Chandra R, Ahmed B, Nimesh S (2016) Advances in preparation and characterization of chitosan nanoparticles for therapeutics. Artif Cells Nanomedicine Biotechnol 44:305–314. https://doi.org/10.3109/21691401.2014.948548
42. El-Naggar NE-A, Shiha AM, Mahrous H, Mohammed ABA (2022) Green synthesis of chitosan nanoparticles, optimization, characterization and antibacterial efficacy against multi drug resistant biofilm-forming Acinetobacter baumannii. Sci Rep 12:19869. https://doi.org/10.1038/s41598-022-24303-5
43. Ramezani F, Moghadasi M, Shamsasenjan K, Narmani A (2024) Folic acid-decorated chitosan-PLGA nanobiopolymers for targeted drug delivery to acute lymphoblastic leukemia cells: in vitro studies. Technol Cancer Res Treat 23. https://doi.org/10.1177/15330338241308077
44. Fatima S, Ikram M, Haider A, Shahzadi A, Moeen S, Ul-Hamid A, Ullah H, Ali G, Salem MA (2025) Dual functionality of chitosan and CTAB doped SnSe nanostructures: RhB decolorization, oxygen evolution reaction and antimicrobial activity against S. aureus by inhibiting DNA gyrase through molecular docking. Int J Biol Macromol 301:140433. https://doi.org/10.1016/j.ijbiomac.2025.140433
45. Waqas M, Shahzadi A, Haider A, Hamid AU, Algaradah MM, Abd-Rabboh HSM, Ikram M (2023) Chitosan grafted polyacrylic acid doped $MnO_2$ nanocomposite an efficient dye degrader and antimicrobial agent. Int J Biol Macromol 251:126343. https://doi.org/10.1016/j.ijbiomac.2023.126343
46. Ainurofiq A, Wiyono N, Vania Febrianti Marzuki A, Choiri S (2025) Screening and assessment of the chitosan-based nanoparticle formation through an ionic gelation method. J Phys Conf Ser 2945:012046. https://doi.org/10.1088/1742-6596/2945/1/012046
47. Athavale R, Sapre N, Rale V, Tongaonkar S, Manna G, Kulkarni A, Shirolkar MM (2022) Tuning the surface charge properties of chitosan nanoparticles. Mater Lett 308:131114. https://doi.org/10.1016/j.matlet.2021.131114
48. Ramírez L, Corral D, Betanzo I, Rodarte D, Chauhan K, Vazquez-Duhalt R (2025) Effect of surface properties of chitosan-based nanoparticles in the skin-diffusion rate. Biopolymers 116:e70006. https://doi.org/10.1002/bip.70006

# Chapter 4
# Bovine Brucellosis

**Abstract** Bovine brucellosis is significant infectious disease caused primarily by *Brucella abortus, Brucella melitensis,* and *Brucella suis,* instigating noteworthy economic losses and public health risks owing to their zoonotic potential. Brucellosis is regarded as most prevalent zoonoses in world by different international organizations. *Brucella* is extremely contagious and transmitted via direct contact by *Brucella* carriers or via indirect contact upon animal's consumption of contaminated excrement or forages carrying abundant bacteria, usually shed via infected animals. *Brucella* replication triggers invasion of inflammatory cells, vasculitis, trophoblastic necrosis, and allantochorion ulceration. Consequently, compromised fetal-maternal metabolic exchanges led to abortion. Traditional diagnostic techniques including culture, serological tests and nucleic acid amplification tests are employed for disease diagnosis. However, these methods present several drawbacks, such as high costs, labor-intensive processes, lower specificity and sensitivity, time consumption, and challenges in differentiating between past and active infections. Next-generation sequencing (NGS) is an innovative diagnostic technology that theoretically has the capacity to detect all microorganisms, including brucellosis, in samples containing DNA as their genetic material. Machine learning (ML) is a contemporary analytical platform that has the potential to improve diagnostic accuracy for brucellosis. Biosecurity measures and vaccination are essential for infection control and the protection of both animal and human health.

**Keyword** Abortion · Bovine brucellosis · NGS · Vaccination · Zoonoses

## 4.1 Introduction

Brucellosis is an infectious, contagious, and zoonotic illness worldwide instigated via bacteria of the genus *Brucella* [1]. It is regarded as the most prevalent zoonosis in the world by different international organizations [2]. Bovine brucellosis is an economically significant disease instigating reproductive failure as the principal manifestation that results in premature birth, abortion, along with lessened milk production



A. Haider et al., *Chitosan Nanoparticles for Brucellosis Treatment*, Springer Series in Biomaterials Science and Engineering 21, https://doi.org/10.1007/978-3-032-16292-2_4

in females, and epididymitis as well as orchitis in males. Disease is caused via facultative intracellular Gram-negative bacterium that has ability to survive within leukocytes and macrophages which protects it from cellular and humoral activities [3].

Brucellosis disease is also known as epizootic abortion, Bang's disease, ram epididymitis, enzootic abortion, slinking of calves, and contagious abortion in animals, while it is named as Malta fever, Mediterranean fever, Undulant fever, and Gastric fever in humans [4]. Presence of brucellosis illness in humans is highly reliant on the presence of disease in animal reservoirs, along with wildlife. Before pasteurization implementation, milk was main infection route in human population. Presently, disease is principally occupational (affects employees in animal industries, abattoirs, and animal health occupations) [5].

Brucellosis ranks first from list of zoonotic bacterial infections, and 500,000 cases recorded each year in endemic regions. Brucellosis is believed to be recognized in the late Roman era, named owing to its similarity with the organism *Brucellae* (later named *Brucella*) observed in carbonized cheese. The disease is linked to military campaigns, mainly in the Mediterranean region. The illness was first explained in detail by Sir David Bruce, Zammit, and Hughes while conducting a study in Malta; thus, the name "Malta fever" is sometimes used for common fever conditions triggered through *Brucella* and its most frequent species as *B. abortus* and *B. melitensis* [6]. *B. abortus*, like other species of the genus *Brucella*, is a facultative intracellular bacterium infecting mammalian hosts. *Brucella* are small, Gram-negative, and coccobacillary bacteria within the class Alphaproteobacteria. *B. abortus* contains two circular chromosomes which encode around 3.2 Mb and is divided into seven biovars based on biochemical, antigenic, and phenotypic properties [7].

*B. abortus* first reported as a causative agent of preterm delivery in cattle and recurrent fever in humans. Bacteria infect the spleen, lymph nodes, and reproductive tissues and consequently cause swelling, edema, and necrosis. It instigates placental lesions in pregnant animals and raises the abortion risk. Brucellosis attracts public health significance when humans acquire infection by unpasteurized meat, milk, and animal byproducts from infected livestock. Accurate diagnosis is a significant barrier in the complete elimination of brucellosis. Though various serological tests such as serum agglutination test, Rose Bengal test (RBT), and enzyme-linked immunosorbent assay (ELISA) are employed for diagnosing disease in cattle; nevertheless, these are frequently observed to be deceptive [6]. *Brucellae* employ numerous mechanisms for protection against DNA mutations, thus there is 94% nucleotide sequence similarity across the genus. Strong homology in the *Brucella* genus led several researchers to suggest that Brucellae represent a single species. Molecular approaches like restriction fragment length polymorphism designed for assessing genetic relationships among isolates are unsuccessful for strain differentiation or epidemiologic examinations due to the stability of the *Brucella* genomes. Variable number tandem repeat is the latest molecular method that assesses strings of the nucleotide repeats in the chromosome's noncoding areas. It has been employed for comparing genetic relationships among strains [7]. PCR techniques as well as serological tests are broadly utilized to confirm appropriate diagnosis in latest years. Apart from the threat to public health, this disease also increases economic challenges to livestock owners and product consumers [6].

Since ancient times, man is closely related with animals which created chance of intercommunicability of the microbial infections between animals and humans. Many cattle diseases are transmitted to humans. Among these, brucellosis is extremely infectious disease of animals and humans that has been documented from various countries of world including India and Ethiopia. Brucellosis in cattle is typically produced via *B. abortus* biovars. In some regions, mainly in western Asia and southern Europe, where cattle are raised in close proximity with goats or sheep, the disease might also be triggered via *B. melitensis. Brucella* primarily affects cattle, goats, sheep, pigs, and camels, as well as initiating disease in humans. Cattle are the primary hosts of *B. abortus*. Humans are vulnerable to both species *B. abortus* and *B. melitensis*, while latter being more commonly reported in humans. It is an economically vital illness of livestock instigating reproductive wastage by infertility, loss of calves, delayed heat, reduced milk and meat production, and culling along with economic losses from global trade bans [8].

*B. melitensis* infection naturally occurs in sheep as well as goats in the Mediterranean region; however, it is widespread globally. Epidemiologically and pathologically, this infection is much similar to *B. abortus* infection in cattle, and major clinical signs are reproductive failure such as unthrifty offspring birth or abortion, epididymitis, and orchitis. Biovars 1 and 3 of *B. melitensis* are most commonly identified in the Mediterranean countries. In most conditions, main sources of excretion include fetal fluids along with vaginal discharges following full-term delivery or abortion. As a result, *B. melitensis* excretion occurs significantly in milk and is commonly found in semen. As observed in cattle, *B. melitensis* infection can be vertically passed from ewes to lambs, though these animals remain affected; however, they are seronegative. Nevertheless, most latent infections are possibly transmitted via milk or colostrum [9].

Brucellosis has high incidence rates in south along with eastern European Union countries as well as in various economically disadvantaged countries. These countries altogether comprise over 70% of world's vulnerable livestock, thus making brucellosis a significant disease worldwide. Infection by *B. melitensis* in goats and sheep is usually neglected, since ruminant production at small scale generally signifies low-income activity carried out by landless farmers in marginal rural regions in the developing world. Owing to marginal and typically migratory farming systems, eliminating and controlling this disease is very difficult. This disease is distributed worldwide; most countries are experiencing re-emergence of infection in goats and sheep and consequently, in humans. The causes of high prevalence might be due to sociocultural factors; however, exacerbated by inadequate control measures adopted in small ruminant farming. Major risk factors for humans infecting with *B. melitensis* are occupational exposure, interaction with animals, food practices, and poor hygienic measures. Since interhuman transmission is uncommon, small ruminants are the leading source for cases of brucellosis in humans. Humans might acquire infection directly via interaction with oronasal or conjunctival mucosa of diseased animals, and indirectly via ingesting tainted animal products, mostly dairy products. At present, there is no feasible technique to prevent human brucellosis, thus for safeguarding people, awareness should be directed toward effectually preventing disease in goats as well as sheep. Though sheep along with goats are main reservoirs for human infection, nevertheless in some countries' buffaloes, bovines, camels, and yaks are also associated. Unfortunately, there is limited knowledge on substitutes to control infection by *B. melitensis* bacteria in such species. Prophylactic and diagnostic tools for brucellosis have been adequately validated for successfully preventing *B. melitensis* infection in goats along with sheep, in some socioeconomic situations. Effectiveness of national veterinary facilities should be improved and administrative organizations must be involved for ensuring success [10].

Brucellosis is one of the main reproductive illnesses in buffaloes triggering abortion storms in the breeding season during the last third of pregnancy, fetal membrane retention, stillbirths, as well as decline in milk yield leading to significant economic losses. Approaches for elimination and control of bovine brucellosis are presently based on animals' detection, restricting animals' transport, prompt detection, and elimination of infected animals by employing various diagnostic tests such as Buffered Acidified Plate Antigen Test (BAPA), RBT, micro-standard agglutination test (mSAT), Complement Fixation Test (CFT), and Rivanol Test (Riv.T.). Additionally, generally at initial stages of the control programs owing to high prevalence, vaccination is administered for avoiding dissemination of the pathogen. *B. abortus* S19 vaccine is the most extensively employed vaccine to prevent brucellosis in cattle [11]. Although practical and scientific knowledge on brucellosis has increased considerably in most countries, brucellosis remains a re-emerging zoonosis, producing noteworthy economic losses [12].

## 4.2 Historical Background

Brucellosis is an ancient illness, possibly originated during the fifth plague in Egypt (approximately 1600 BC). Latest investigation of antique Egyptian bones from nearly 750 BC demonstrated evidence of osteoarticular and sacroiliitis lesions that are general complications of the disease [13]. Brucellosis occurrence was stated in the Mediterranean region and linked with war campaigns historically [14]. In 1887, David Bruce identified *B. melitensis* (then named *Micrococcus melitensis*) from a British soldier's spleen who passed away from Malta fever, a febrile illness that was commonly present among military members stationed in Malta. For nearly twenty years after *M. melitensis* isolation, Malta fever persisted as a mystery and was believed to be a vector-borne illness till Themistocles Zammit inadvertently confirmed the zoonotic nature of brucellosis in 1905 via isolation of the *B. melitensis* pathogen from the milk of a goat. It was thought that goats are not an infection source as they demonstrated no signs of disease upon inoculation with *Brucella* cultures. The finding that healthy goats can serve as disease carriers has been regarded as the greatest advancement made in epidemiology research. Bernhard Bang, a Danish veterinarian, in 1897 identified cattle abortion bacillus or Bang's bacillus (*B. abortus*) as the etiologic agent of Bang's illness [13]. Huddleson and Traum isolated *B. suis* from swine that is documented to instigate human brucellosis [14]. An American scientist, Alice Evans, conducted landmark research on virulent bacteria in dairy products, proved the relationship between Malta fever and Bang's disease, and renamed the genus *Brucella* to honor David Bruce. Her study on *Brucella* was significant in accepting the pasteurization process for preventing human brucellosis in the USA [13]. Evans discovered that *B. melitensis* (*Micrococcus melitensis*) recovered from pigs and cows is classified within the same genus. Stoenner and Lackman recovered *B. neotomae* from rats. *B. canis* isolated from dogs by Bruner and Carmicheal. *B. ceti* and *B. pinnipedialis* are relatively recent *Brucellae* identified during the last decade from marine mammals and could pose a zoonotic concern in the future. *B. microti* reported in terrestrial animals [14]. Discovery of *Brucella* in early 1990 in marine mammals has altered the notion of land-based brucellosis distribution and related control procedures [13]. The isolation of different *Brucella* strains from human beings as well as marine mammals lately reveals the importance of zoonotic transmission [14].

## 4.3 Etiology and Characteristics

### *4.3.1 Etiology of Brucellosis*

Human brucellosis is a zoonosis with a wide variety of clinical symptoms, instigated by small, aerobic, and Gram-negative rods of genus *Brucella* [15]. *Brucella* are small (0.5–0.7 μm wide and 0.6–1.5 μm long), non-motile, Gram-negative, non-spore forming, non-encapsulated, coccobacilli (rod-shaped) bacteria that are facultative intracellular parasites (Fig. 4.1). The genus demonstrates slight variation genetically [16]. After penetrating the body, bacilli multiply in the lymphatic system, primarily in lymph nodes, consequently cross the protective barrier and spread to various organs. Genus *Brucella* was discovered in 1887 by David Bruce and currently comprises ten species (with many serotypes). These are:

1. ***Brucella melitensis—identified*** in Malta in 1887 by David Bruce from the spleen of a soldier who passed away due to acute brucellosis. This species is most virulent for humans.
2. ***Brucella abortus—triggering*** cattle abortions, and main causative agent of brucellosis in humans (Bang's disease) along with animals for several years in Poland.
3. ***Brucella suis—instigating*** infection mostly in swine and is also pathogenic for humans. It was also recovered from wild hares in Poland.
4. ***Brucella canis—recovered*** from dogs—can also instigate disease in humans. It was described initially in 1966 by Carmichael, who recovered bacillus from fetuses, placenta, and vaginal discharge from bitches who aborted their litters. Disease was identified earlier in Beagle dogs in the United States.
5. ***Brucella neotomae—isolated*** from rats in United States.

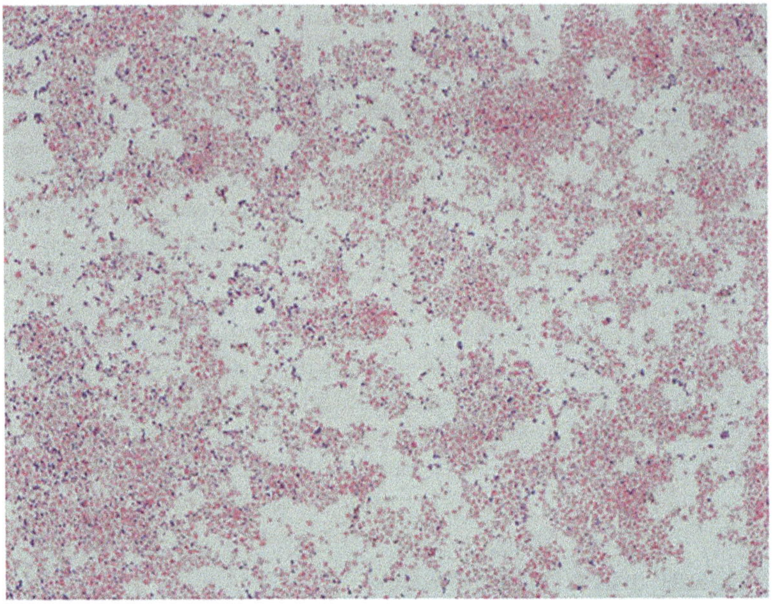

**Fig. 4.1** Gram stain from bone marrow culture demonstrated small Gram-negative coccobacilli. This was later confirmed as *B. melitensis*. Reproduced with permission from Ref. [18]

6. ***Brucella ovis—cause*** infection in sheep as well as cultured from semen of rams in Poland.
7. ***Brucella marina–Brucella****ceti* is present in sea mammals such as seals and whales in the Atlantic Ocean. *Brucella pinnipedialis* is also discovered in sea mammals [15].
8. ***Brucella papionis—recovered*** from cervical swab collected from baboon after stillbirth at handling facility of animals in Texas, USA in 2006 [17].
9. ***Brucella microti—identified*** from common vole in Czech Republic, also from soil in similar region after several years as well as from mandibular lymph nodes in red wild foxes in Austria.
10. ***Brucella inopinata—recovered*** from woman's breast implant wound with clinical symptoms of brucellosis.

Above-mentioned *Brucella* species are pathogenic for different mammals, as wild animals including breeding stock, small rodents, and domesticated animals, along with birds and sea mammals. Breeding stock such as cattle, goats, sheep, dogs, and poultry, as well as wild animals like hares, rodents, etc., are both vector and reservoir of disease in humans [15].

*B. abortus*, *B. suis,* and *B. melitensis* are main causative agents of bovine brucellosis. *B. abortus* is a small, facultative intracellular, and Gram-negative bacterium [19]. Some *Brucella* species are divided further into different biovars. *B. abortus* is subdivided into biovars 1, 2, 3, 4, 5, 6, 7, and 9. *B. melitensis* biovars 1, 2, and 3 as well as *B. suis* biovars 1, 2, 3, 4, and 5 are also reported. Biovar 1 of *B. abortus* accounts for over 80% of total isolates worldwide while in Africa biovar 3 of *B. abortus* reported mostly in a few published studies [16].

### *4.3.2 Characteristics of* **Brucella**

Genus *Brucella* classified under α-2 subdivision of Proteobacteria are Gram-negative eubacteria with cell envelopes composed of outer and inner membranes enclosing periplasm with peptidoglycan mesh as well as soluble components [20]. *Brucella* are small, faintly stained coccoid rods, appear like "fine sand" under microscope lacking capsules, endospores, and natural plasmids. Principal *Brucella* culture displays non-pigmented, punctate and non-hemolytic colonies while colonies of smooth strains of *Brucella* are convex, raised, translucent and circular. After long-term culture of more than 4 days or sub-cultivation, colonies of *Brucella* become less convex, opaque with dry, dull, yellowish-white, and demonstrated granules-like appearance. These alterations instigated by *Brucella* dissociation from smooth to rough forms. When *Brucella* species are stained with safranin, they exhibit weak staining [21]. Historically, species classification of genus *Brucella* is based on host preference combination along with phenotypic biotyping, which assesses variety of characteristics like dye-sensitivity, $H_2S$ production, $CO_2$ requirement, agglutination with the monospecific sera and lysis via *Brucella*-specific bacteriophage. Phenotypic analysis is carried out by characterization of growth on various media, $H_2S$ production, $CO_2$ requirement, growth in dyes presence like basic fuchsin and thionin, reaction with unabsorbed *Brucella* and monospecific A as well as M antisera along with microscopy [17]. Though *Brucella* is obligate aerobe, certain strains need $CO_2$, particularly on primary isolation. *Brucella* is usually urease-positive and oxidase-positive.

*Brucella* is an intracellular pathogen; it persists and replicates in macrophages during infection. Bacteria adapt to low oxygen level, acidic pH, and low nutrient levels. Bacterial cells can withstand long periods in water, meat, soil, aborted fetus, dairy products, dust, and dung. For isolating *Brucella species*, enrichment as well as selective media like Farrell's medium or Tayer-Martin's medium are commonly employed. The colonies become mature after 4–6 days of incubation at 37 °C temperature. They might also grow slowly and poorly at 28 °C. Furthermore, *Brucella* could grow in both 10% $CO_2$ and aerobic atmosphere, while growth is improved without further carbon dioxide on serum dextrose agar. Biphase medium known as Castaneda must be used for blood or body fluid culture. Castaneda contains 2 phases: solid and liquid in a closed bottle. Liquid medium consists of 1–2% sodium citrate. A sample of 5–10 ml is incorporated in medium and then incubated at 37 °C in a perpendicular standing bottle containing 10% $CO_2$ atmosphere. Bone marrow culture might have higher yield than standard cultures of blood [21].

## 4.4 Host Range

There are ten *Brucella* species on the basis of preferred host specificity: *B. abortus, B. melitensis, B. canis, B. suis, B. neotomae, B. ovis, B. pinnipedia, B. cetacea, B. inopinata*, and *B. microti* (Table 4.1). From 10 recognized *Brucella* species, *B. melitensis, B. canis, B. abortus*, and *B. suis* are harmful for humans. Human infections owing to *B. canis* are rare. *B. abortus, B. suis*, and *B. melitensis* are highly pathogenic for humans. These bacteria recognized as favorable agents for bioterrorism and categorized by the US Center for Disease Control (CDC) as B category priority pathogens [22]. Even though all these four species are able to instigate human brucellosis, *B. melitensis* is highly common globally and is the main etiologic agent of brucellosis in humans [23].

Bovine brucellosis is predominantly instigated by *B. abortus*, specie causing infection in water buffaloes, camels, yaks, and infrequently in small ruminants, dogs, pigs, reindeer, horses, and few wild animals. Water buffalo is widespread domestic animal and major locus for brucellosis infection. Despite water buffaloes mainly thrive in marshy areas in humid tropical areas and rivers, however they are also raised on inhabited continents as livestock. They are not only employed to reclaim farmland, but their production of meat, milk, and leather is of economic significance. Brucellosis in cattle is mostly instigated by *B. abortus* triggering infertility and abortion in adult animals; however, infection can also occur via *B. melitensis,* most frequent agent of brucellosis in goats and sheep. This infection does not eliminate naturally upon removing infected small ruminants. Cattle can rarely infect by *B. suis*, it is typical agent causing swine brucellosis [24–26]. Sheep and goats are primary hosts of*B. melitensis* that is considered most virulent among different *Brucella* species [27].

*B. suis* biovars 1–3 instigate brucellosis in domestic swine and to limited extent, either by *B. melitensis* or *B. abortus*. Geographic distribution, host range, and pathogenicity of *B. suis* vary based on biovars. *B. suis* biovars 1 and 3 are zoonotic,

**Table 4.1** Zoonotic potential and host preference of *Brucella* species

| Species | Zoonotic potential | Host preference |
|---|---|---|
| *B. melitensis* | High | Sheep, goat |
| *B. abortus* | Moderate | Cattle |
| *B. suis* | Moderate | Pig |
| *B. canis* | Mild | Dog |
| *B. ovis* | Absent | Sheep |
| *B. neotomae* | Absent | Desert wood rat (*Neotomae lepida)* |
| *B. ceti* | Mild | Cetaceans |
| *B. pinnipedialis* | Mild | Seals |
| *B. microti* | Absent | Common voles (*Microtus arvalis*) |

Reproduced with permission from Ref. [33]

as well as endemically found in Americas, Oceania, Asia and infrequently in Europe; affect various animal species, mainly *Suidae* family, however also humans, horses, cattle, and dogs. *B. suis* biovar 2 is confined to Europe and it has minor zoonotic risk, though it endangers pig production in Europe and cause infection in cattle. European hare and Wild boar are reservoirs for *B. suis* biovar 2 and appear as relevant in transmission of disease to farm pigs. *B. suis* biovar 4 infects reindeer, arctic fox, arctic wolf, cattle (without instigating disease), and humans. *B. suis* biovar 5 infects rodents [28].

Canine brucellosis is triggered by *B. canis* bacterium. *B. canis* is a zoonotic pathogen, can infect humans by direct contact with excreta and secretions of infected dogs or via laboratory exposure [29]. *B. ovis* is a naturally rough bacterium that belongs to *Brucella* genus and causes ovine epididymitis. *B. ovis* infection is documented in sheep farming globally but to date, the real distribution of this infection in the world is highly unknown. *B. ovis* infection does not show clear, pathognomonic clinical symptoms; often illness is subclinical and might circulate in flock without suspicion [30]. *B. neotomae* is a smooth strain, recovered from desert woodrats. It has been demonstrated that *B. neotomae* might also instigate human infections. Furthermore, *B. neotomae* attacks spleen, liver, lymph nodes, and causes Th1 cytokine expression in mice like *B. melitensis*, *B. suis*, and *B. abortus*. Consequently, *B. neotomae* signifies a possible zoonotic species [31]. *B. pinnipedialis* and *B. ceti* are isolated from lungworms as well as a variety of organs from marine mammals. There are numerous reports on isolation of *B. ceti* along Spanish and Italian Mediterranean coasts from bottlenose and striped dolphins [32].

A new *Brucella* species recovered initially from common vole during an epizootic brucellosis outbreak in the vole population of southern Moravia. *B. microti* was identified in the Czech Republic from soil, mandibular lymph nodes from red foxes in Austria, as well as wild boar in Hungary. Lately, *B. microti*-like strains were isolated from indigenous moor frog in a French farm. Since *B. microti* detected in soil, rodents, and *B. microti*-like strains isolated from farmed frogs, knowledge of possible public health threat must be highlighted, particularly among extremely exposed groups such as hunters, farmers, zoologists, forestry workers, veterinarians, and laboratory technicians [34]. Recently, novel spp. *B. inopinata* with type strain $BO1^T$, recovered from blood and breast implant wound fluid of a female patient (71 y/o) with clinical symptoms of brucellosis [35].

## 4.5 Transmission

*Brucella* is extremely contagious and transmitted via direct contact by *Brucella* carriers or via indirect contact upon animals' consumption of contaminated excrement or forages carrying abundant bacteria, usually shed via infected animals (Fig. 4.2). *Brucella* can persist in soil for 20–120 days, in water for 70–150 days as well as two months in food such as meat and milk [36]. *B. abortus* can infect several hosts. The main transmission modes for *B. abortus* in cattle are via direct contact with vaginal secretions of animals which are infected or their aborted products [37]. Because dairy cows acquire infection mostly by consumption of contaminated excrement or forages discharged through infected animals, infection via environment is crucial, even more significant than direct transmission via contact among individuals [36]. Interacting with wild animals, for instance, deer and rodents, and with ticks also contribute to recurrence of infection by *B. abortus* in domestic animals. Physical contact with animals that are infected and aborted products of animals without using personal protective equipment (PPE) such as masks and gloves is regarded as a hazardous way of acquiring brucellosis in humans. Consuming raw as well as unpasteurized products in dairy along with inhalation of *Brucella* aerosols are usual routes of *Brucella* species contraction in humans. Thus, *B. abortus* is considered a One Health hazard for animals, occupational workers, and public [37].

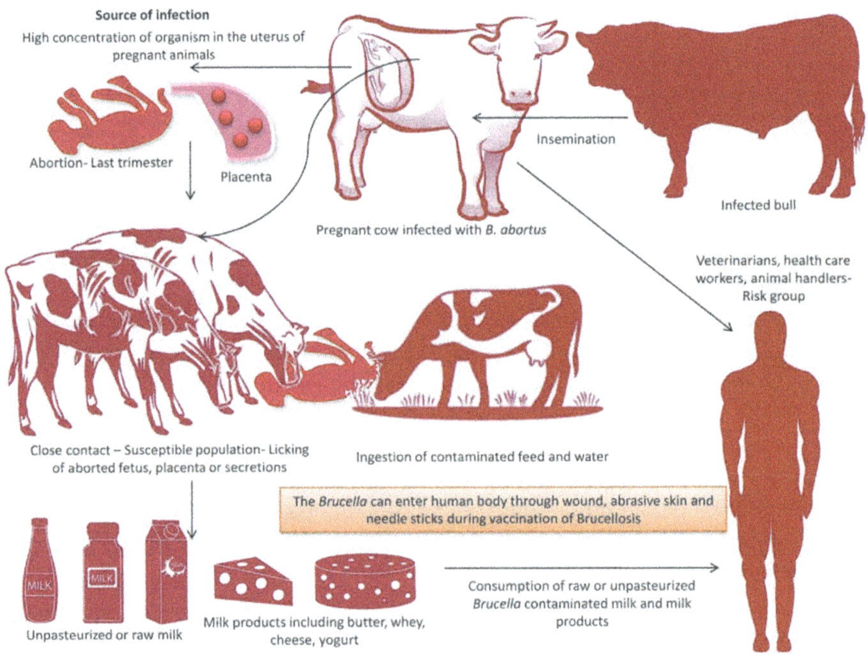


**Fig. 4.2** Transmission of Brucellosis. Pregnant cows usually abort in the last trimester of pregnancy. Aborted fetus, placenta, and secretion from uterus act as the source of infection to other animals. Milk and milk products can act as source of infection to man if consumed unpasteurized. Infected bulls serve as the lifelong source of infection. Reproduced with permission from Ref. [14]

## 4.6 Geographical Distribution

Animal and human brucellosis have been reported from over 170 countries as well as regions worldwide. Each year, about 500,000 instances of brucellosis in humans are documented. Brucellosis is endemic in several countries, and incidence ranges from < 0.01 to > 200 cases per 100,000 population. Notification rate in EU/EEA reported 0.04 cases per one hundred thousand people. Greece shows highest notification rate of 0.33 cases per 100,000 [38]. Millions of fresh human brucellosis cases are reported annually on a global basis [39]. This disease is yet endemic in Mediterranean, Central Asian, and Middle East countries. In some North American and European countries, disease has been eliminated due to restrictions which are posed on the international trade of animals along with animal products. As compared to North America, Europe, and additional developed regions of world, greatest impact of infection is reported in Mediterranean, South as well as Central America, Asia, Africa, Eastern Europe, Indian subcontinent, and Middle East, especially Syria, Iraq, Turkey, Egypt, and Iran. Incidence of this illness varies significantly among countries. Egypt and Iraq have observed 4–fivefold increase relative to other regions. Iran ranks second worldwide in prevalence of brucellosis. It is endemic in certain parts of Iran. Recent statistics demonstrate that highest prevalence in Iran, with 31–41 cases per 100,000 people, is observed in provinces of Markazi, Lorestan, Kermanshah, and East Azarbaijan [40]. Many countries were successful in eliminating brucellosis worldwide from their livestock like Australia and certain EU member states such as Spain and Croatia where disease is almost wiped out, while in Southern Europe as Greece, Italy, and Portugal, infection in livestock and human zoonotic cases still alert public health authorities as well as veterinary services [41].

## 4.7 Pathogenesis

Aborted fetuses and fetal membranes as well as uterine secretions removed after parturition or abortion are the most significant infection sources. Infection can also spread to calves vertically along with contaminated milk; nevertheless, these infection routes are far less significant. Venereal transmission under natural conditions is not a major infection route; however, artificial insemination with tainted semen is a potent infection source. Though infection can be acquired via respiratory mucosa through inhalation, conjunctiva and skin while gastrointestinal tract is most frequent infection route in cattle, from where infection disseminates to local lymph nodes, there *Brucella* multiplies within cell in phagocytes. Bacteremia causes systemic infection after lymphatic vessel invasion, favoring colonization of pregnant uterus, mammary gland, and male genital organs. *B. abortus* exhibits strong preference for the uterus during the final gestation trimester that is believed to be owing to high concentrations of steroid hormones and erythritol. Erythritol promotes survival of bacteria because it is metabolized through *B. abortus* as an energy and carbon source. *B. abortus* propagates to intercotyledonary trophoblasts from erythrophagocytic trophoblastic cells present at the chorionic villi base of ruminants, which are thought to be the primary invasion site of fetal placental tissues. *Brucella* replication triggers invasion of inflammatory cells, vasculitis, trophoblastic necrosis, and allantochorion ulceration. As a result, fetal-maternal metabolic exchanges are compromised, instigating abortion (Fig. 4.3) [42].

The mechanisms allowing host cell invasion via *Brucella* spp. remain partially understood. Though particular host receptors which interact with *Brucella* are not recognized so far, *Brucella* internalization in host cells demands cytoskeletal changes. Notably, invasion via digestive tract doesn't induce inflammatory response of host. Thus, *Brucella* species attack stealthily or undetected by host's innate immune system. In fact, *Brucella* species possess mechanisms which avoid host's innate immune-system activation. Indeed, *Brucella* Toll/interleukin-1 receptor (TIR) domain consisting of protein inhibits signaling of Toll like receptor 2 (TLR2) via inhibition of MyD88, DC maturation, cytokine secretion and consequently, antigen presentation. *B. abortus* also inhibits pro-inflammatory mediator's transcription in trophoblastic cells at initial infection stages. Trophoblasts are cells of placenta which are targeted throughout infection in pregnant cows. *B. abortus* causes expression of the pro-inflammatory chemokines through cultured trophoblastic cells after initial suppression of the pro-inflammatory transcripts, which correlates with expression profile observed in-vivo in placenta of infection positive cows.

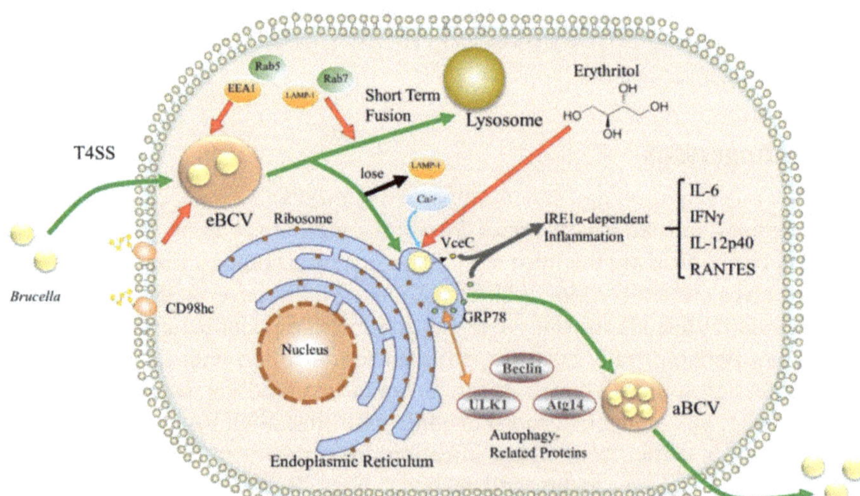


**Fig. 4.3** Intracellular process of *Brucella* invading CTB. Green arrows indicate the infection pathway of *Brucella*; red arrows indicate adsorption, acquisition or capture; gray arrows indicate co-cause and mediation; orange arrows indicate interactions; blue arrows indicate binding; black arrows indicate loss. Reproduced with permission from Ref. [43]

*Brucella* spp. lacks traditional bacterial virulence factors such as capsule, cytolysins, fimbriae, flagella, exotoxins, lysogenic phages, endotoxic lipopolysaccharide (LPS), plasmids, and mediators of host cell apoptosis. Nevertheless, LPS plays a significant role in virulence of *Brucella* because it precludes complement-facilitated eradication of bacteria and confers resistance against the antimicrobial peptides such as lactoferrin and defensins. Another significant mechanism of virulence is BvrR/BvrS two-component regulatory system that is necessary for the modulation of cytoskeleton of host cell during *Brucella* invasion as well as for regulating outer membrane proteins expression, certain of which are essential for complete virulence. Cyclic β-1,2-glucans (CβG), that are outer membrane part, are also necessary for *Brucella* intracellular survival. *Brucella* species secrete a type-IV secretion system (T4SS), that is encoded by *virB* operon components, imperative for intracellular survival of bacteria in host cells, along with in vivo virulence. Transcription of *virB* operon is prompted particularly inside macrophages and phagosome acidification is a significant intracellular signal causing *virB* expression as compared to other type-IV systems, that are expressed extracellularly. Placentitis prevents nutrient delivery to fetus leading to fetal stress along with death, and it has been hypothesized as a reason for abortion in brucellosis (Fig. 4.4). Researchers have suggested that hormonal influences facilitate the abortion process. Shift from predominant production of progesterone to estrogen instigates endometrial production of prostaglandin F 2-alpha (PGF2α), that induces premature or normal delivery process [44].

## 4.8 Clinical Signs and Symptoms

The main clinical sign is abortion in pregnant females in bovines (buffaloes and cows). Abortion typically occurs from 5 to 8th gestation month. The incidence of abortion is associated with some factors, such as stage of pregnancy, number of organisms infecting, and animal resistance. Instead of abortion, stillborn, premature, or weak calves can also occur. Abortion is frequently followed by placental retention along with metritis, which can cause transient or permanent infertility. Brucellosis doesn't usually cause gross organic lesions; it can sometimes instigate a slight interstitial inflammatory response in the mammary gland, that is associated with eradication of bacteria from milk. Macrophages of the mammary gland might provide an intracellular environment for *B. abortus* persistence in the mammary gland of infected cows. Although being susceptible to infection, infected heifers which are at the prepubescent stage have not demonstrated any clinical signs. There is proof that heifers which are infected before adulthood might abort and contribute to maintaining infection in the herd [45].

*B. abortus* mainly instigates illness in cows, can be recovered from udder, lymphoid organs, and uterus. Brucellosis outbreaks in dairy herds lead to decrease in milk production, milk somatic cell count increases, abortions occurrence, and postpartum metritis. Late abortion is related to necrohemorrhagic placentitis as well as fetal lesions, predominantly fibrinous pleuritis, pneumonia, and pericarditis. Infected

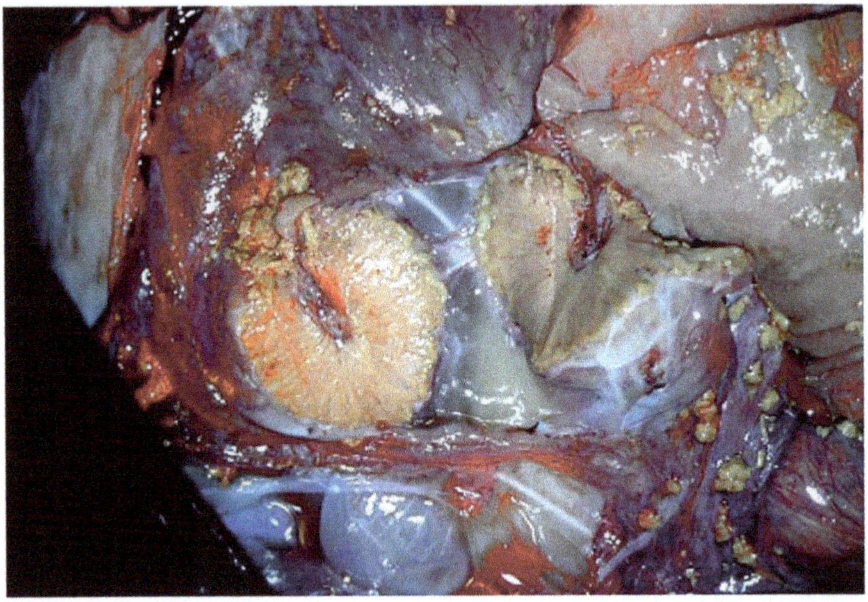

**Fig. 4.4** Cow experimentally infected with *Brucella abortus*. Uterus containing multifocal fibrinous exudate on the caruncular surface. Cut surface of a placentome with fibrinous necrotic exudate and multifocal hemorrhage (necrotizing placentitis). Reproduced with permission from Ref. [42]. Copyright 2010, Elsevier

cows generally abort only once, and then gestations can produce calves that might be born either healthy or weak. Several infected cows do not display any clinical signs of disease and may generate healthy calves. Transmission occurs mostly after parturition or abortion of infected cows by contaminated fetus, uterine secretions, and fetal membranes. Infection also occurs in bulls; however, they may not quickly transmit disease. *B. abortus* is general reason of orchitis, frequently related to epididymitis and vesiculitis. Depending on intensity of lesions, infection in males causes either permanent or temporary infertility [46].

Brucellosis in humans is distinguished by varied incubation periods ranging from some days to various months. It is a subacute or acute febrile infection and complication that affects systems and organs. Brucellosis common symptoms are unclear and analogous to flu. These symptoms are intermittent, irregular, or continued fever of varying duration with weakness, headaches, depression, profuse sweating, chills, abdominal pain, cough, and weight loss. Localized suppurative infections might also happen. Abortion can occur within initial trimesters of pregnancy [23].

## 4.9 Zoonotic Potential

Human brucellosis is zoonotic disease with significant impact on public health, despite successful eradication as well as control programs for domestic animals established in various countries around world [47]. *Brucella* can cause infection in numerous animal species and presently contain twelve well-classified species, from which 4 are instigating nearly entire human infections as *B. abortus*, *B. melitensis*, *B. canis,* and *B. suis* with *B. melitensis* exhibiting high virulence [48]. Infection is generally transmitted from animal reservoir to humans either via direct contact or by indirectly through contaminated foods like unpasteurized cheese or milk. The disease mainly presents as multiple clinical signs along with symptoms including fever of unknown origin [47]. Although clinical presentations of disease are variable, most usual clinical sign is the undulant fever in which temperature vary from 37 °C in morning to 40 °C in afternoon. Other symptoms are night sweats with peculiar odor, arthralgia, chills, weakness and myalgia. Furthermore, insomnia, malaise, anorexia, constipation, headache, nervousness, depression, and sexual impotence are also general manifestations. It is stated that brucellosis can instigate spontaneous abortion, neonatal and congenital infections in humans. It is demonstrated by some instances in which *Brucella* spp. recovered from placental or fetal tissues. Complications such as arthritis, orchitis, endocarditis, spondylitis, and prostatitis can occur when internal organs are involved. Neurological complications like encephalitis, meningitis, chorea, brain abscess, meningo-encephalitis, facial palsy, ischemic attacks, and meningomyeloencephalo-spondylosis can also arise during onset of disease, recuperation period or even after recovery from acute infection [49]. Chronic disease, primary antibiotic treatment failure along with relapses are frequent [47].

Although using unpasteurized milk as well as milk products raises risk of brucellosis incidence, this is not only route for possible infection. Aerogene route is also significant for spreading bacteria. Brucellosis is one of the potential laboratory infections. Infection occurs owing to working in an unprotected environment by skin, eyes, and mouth, not employing gloves, protecting glasses, mouth masks, via mouth pipetting; these are ways of exposing eyes, mouth, or nose to contaminated aerosol [50]. Though illness might be chronic and debilitating, brucellosis is rarely fatal in humans, excluding some untreated cases because of vegetative endocarditis. Prolonged illness courses, often related with the suppurative destructive lesions, are due to *B. suis* infections. Mild-to-moderate sporadic illness instigated by *B. abortus* seldom causes complications. *B. canis* infection has disease course identical to *B. abortus* infection. Infection due to *B. canis* generally has insidious onset instigating frequent relapses and doesn't normally induce chronic illness in humans. *B. cetaceae* and *B. pinnipediae* lately observed to instigate primarily neurobrucellosis in humans [49]. Global resurgence of *B. melitensis* led to rising prevalence of human brucellosis [51]. Endocarditis triggered by *B. melitensis* accounts for 0.2–0.3% complications from brucellosis of humans, with aortic valve being mostly affected in 75% cases, with 80% estimated mortality rate for all cases [52].

Brucellosis is one of "seven neglected endemic zoonoses" as reported by World Health Organization (WHO) [53]. Though over 0.5 million human instances are stated worldwide annually, the number of unnoticed cases is thought to be significantly higher. This critical situation might be attributed to non-specific clinical symptoms of human brucellosis, little awareness of disease in non-endemic nations, and inadequacies in laboratory diagnosis. Number of human instances is directly connected with number of infected animals in defined regions. Effective countermeasures for reducing human brucellosis incidence are consequently based on surveillance along with livestock control as well as cooking or pasteurization of tainted food products. Once disease is transmitted to humans from animal reservoir, only timely diagnosis and appropriate antibiotic therapy might prevent patients from serious sequelae [47].

## 4.10 Diagnosis

Serological tests are the basis to diagnose bovine brucellosis, because clinical samples that are used are easily obtainable and tests are comparatively inexpensive and simple to perform. However, there are many serological tests utilized for diagnosing brucellosis; there is no test that is completely accurate, and no ideal test notion when utilized alone owing to variations in diagnostic specificity (DSp) and diagnostic sensitivity (DSe), that might be associated to test as well as disease epidemiology. Traditionally, bovine brucellosis is diagnosed serologically by employing a series of confirmatory tests along with screening. Screening tests are extremely rapid, sensitive, and inexpensive such as RBT, indirect enzyme-linked immunosorbent assay (ELISA), and buffered plate agglutination test (BPAT); though, in these tests positive results must be subjected to the confirmatory tests [54]. The BP26 (OMP 28) protein used as antigen in ELISA with analytical specificity and sensitivity of 94% (47/50) and 100% (25/25), respectively, indicating potential of antigen for bovine brucellosis diagnostic test. Swift lateral flow immunochromatographic assays (LFIA) have been developed so far and evaluated worldwide based on *B. abortus* smooth lipopolysaccharide (S-LPS) employment as antigen, with identical sensitivity along with specificity to traditional serological tests for diagnosing bovine brucellosis [55]. Conversely, confirmatory tests such as complement fixation test as well as fluorescence polarization assay have high specificity and good sensitivity; however, these tests normally require complex equipment along with skillful result interpretation [54].

Definitive diagnosis is completely dependent on combined employment of serological and bacteriological methods. Assembled diagnostic method is essential owing to prevailing restrictions of these procedures. Although isolation along with identification of bacteria leads to definitive diagnosis, testing huge herds is rather impractical. Serological approaches are relatively swifter, safer, and easier, thus effectively utilized for mass screening infectious bacteria. However, cross-reactivity of serological processes with other microorganisms which share similar S-LPS frequently result in false positive diagnosis, and delayed antibodies production in primary infection can cause misdiagnosis of disease. To address challenges related to bacteria as well as serological identification, numerous molecular diagnostic assays are developed for detecting *B. abortus*. Polymerase chain reaction (PCR)-based assays are most dominant technology. These techniques have noteworthy advantages over conventional approaches and become valuable tools for utilization in disease diagnostics. Nevertheless, these diagnostic tools must show high diagnostic accuracy (specificity and sensitivity) for supporting accurate and early identification of bovine brucellosis and perhaps overtake serological tests utilization at point-of-care. With increasing use in detecting infection-causing pathogens, careful assessment of such diagnostic assays against the gold-standard techniques can precede application of these techniques on large scale [56].

Owing to non-specific and variable clinical indications in humans, microbiological examination is essential for detection and subsequent management of human instances. Laboratory diagnosis is carried out by employing three distinct approaches or microbiological processes: direct diagnosis via culture, serological tests for indirect diagnosis, and quick diagnosis through PCR-based molecular approaches.

### *4.10.1 Culture*

Culture is regarded as the "gold standard" to diagnose brucellosis in the laboratory. As pathogenesis of human brucellosis is always characterized through an early bacteremic phase, upon suspicion of brucellosis, peripheral blood cultures should be conducted. This represents an important method for confirming disease, even though it shows sensitivity in the range of 10–90%. In certain instances, WHO and the American Society for Microbiology recommend extending BC bottles and blinded subcultures incubation for up to four weeks; however, this protocol has various drawbacks such as high cost, labor-intensive work, diagnostic delays from the automatic BC systems, and organizational challenges.

### *4.10.2 Serological Tests*

Indirect diagnosis based on serological testing is designed for identifying particular antibodies in the patient's serum. Interpretive criteria are consistently described as a high-specificity titer in agglutination tests, cutoff value in ELISA, or existence of a clearly visible certain band in lateral flow immunoassay. Main drawbacks of serological testing are less specificity, difficulty in interpreting results, particularly in patients with recurrent exposure to *Brucella* as well as differentiating between past and active infection. Serology also displayed low sensitivity in initial stages of illness along with suboptimal specificity owing to cross-reaction with further species of bacteria. Additional major reasons for false-negative results are low-affinity antibodies, prozone effect, and *B. canis* infections. Despite these restrictions, serological tests are the main diagnostic techniques for diagnosing brucellosis in endemic, along with low-to-middle-income countries, due to their low cost, user friendly and high-negative predictive value. Wide range of serological tests are available for diagnosing brucellosis in humans, frequently used in the veterinary field. Most serological tests employed for diagnosing brucellosis in the laboratory are categorized into processes that target smooth LPS and cytosolic proteins of *Brucella.* Approaches directed at S-LPS include RBT, standard agglutination test (SAT) in tubes, the most frequent serological procedure employed to diagnose *B. abortus, B. suis, and B. melitensis* infections, along with the latest created SAT miniaturized test.

### *4.10.3 Nucleic Acid Amplification Tests (NAATs)*

Molecular technique known as NAATs facilitated diagnosing brucellosis with high specificity and sensitivity in few hours. Particular targets employed in molecular testing mostly involve genes that encode proteins of outer membrane such as *omp2, omp28* (*bp26*), *and omp31* genes. Additional targets of gene employed for molecular analysis of *Brucella* infection include:

(i) 16S rRNA gene, albeit cross-reactions occurred.
(ii) IS*711* insertion sequence, in which performance is limited owing to variation in sequence along with its non-existence in several strains.
(iii) *bcsp31*, commonly utilized gene, encodes production of immunogenic membrane protein.

Amplification techniques in NAATs are traditional PCR procedures, nested PCR, in-house PCR, quantitative RT-PCR, real-time PCR, PCR-enzyme immunoassay (PCR-EIA) in microplate format, and multiplex real-time PCR (M-RT-PCR) [48].

Next-generation sequencing (NGS) is a novel diagnostic technology employed in the field of pathogenic infectious disease diagnosis. NGS can correctly detect pathogenic bacterial infections via comparing sequences of nucleic acid of the clinical samples with all known microorganism's genomes in short time period through automated bioinformatic analysis and high-throughput sequencing. NGS offers benefits such as high throughput, precision, and rapidity in pathogen diagnosis. NGS overcomes the shortcomings of primer designing for targeted diagnosis in current sequencing and makes available information on virulence genes and resistance mutations, which helps in treatment and clinical diagnosis. NGS theoretically exhibits capability for detection of all microorganisms including brucellosis in samples containing DNA as their genetic material. NGS assay significantly enhances the positive detection rate of brucellosis and offers improved diagnostic effectiveness for both non-brucellosis and brucellosis infectious diseases of joint and bone [57].

Artificial intelligence (AI) has transformed various industries, including the veterinary medicine field. AI has confirmed enormous potential in clinical examining, diagnosing, and treating animals owing to evolving capabilities and usages. AI can detect patterns as well as make predictions which support veterinarians to diagnose complex diseases [58]. Machine learning (ML) is a fundamental approach in AI that permits computers to analyze data and forecast outcomes without requiring explicit programming. ML comprises two main categories: traditional ML and deep learning (DL) [59]. A study on bovine brucellosis applied ultraviolet–visible spectroscopy (UV–vis) by combining it with machine learning (ML) algorithms successfully for disease detection, representing the promise of spectroscopic techniques in diagnosis. Another study revealed that the combination of Fourier-transform infrared spectrophotometry (FTIR) and machine learning allows accurate identification of serum samples from animals infected and non-infected with bovine brucellosis. In dried serum samples, the technique achieved 95.8% accuracy, 100% specificity, and 91.7% sensitivity while for liquid serum samples, the method obtained 91.7% accuracy, 83.3% specificity, and 100% sensitivity. Remarkably, both approaches surpassed conventional gold-standard diagnostic tests [60].

## 4.11 Treatment

Antibiotic treatment in domestic animals for brucellosis is frequently unsuccessful due to intracellular survival and adaptability of *Brucella* in macrophages [14]. Vaccinating animals which are potential bacteria carriers, especially in livestock populations, is favorable for controlling brucellosis distribution. Treatment approaches generally include combination of many antibiotics; nevertheless, antibiotic-resistant *Brucella* strains are emerging and posing significant complications. After treatment, relapse incidence ranges from 5 to 15%. Typically, relapses occur within 6 months following treatment completion, even though they might be observed for up to 12 months subsequently. Rifampin and doxycycline are commonly employed antibiotics for brucellosis treatment. They serve as basis for treatment of all kinds of brucellosis in humans (Fig. 4.5). After appropriate antibiotic therapy, complete recovery is anticipated in non-complicated or acute brucellosis. Children older than 8 years and adults typically take doxycycline orally for 6 weeks, that is the preferred antibiotic owing to dosing frequency as well as less gastrointestinal side effects. To minimize relapse risk, aminoglycosides are frequently added during initial therapy of 2–3 weeks. Fluoroquinolones are regarded as secondary alternatives owing to high efficacy.

Short-term therapy is discouraged due to high relapse rates and failure. For treating complicated cases as spondylitis, endocarditis or neurobrucellosis long-term triple therapy regimen comprising of streptomycin, rifampicin, gentamicin, and doxycycline is more efficacious. Different combinations of oral drugs such as tetracycline/rifampicin, ciprofloxacin/rifampicin, or doxycycline/ofloxacin can be utilized. Rifampicin should never be used alone for avoiding multidrug-resistant tuberculosis

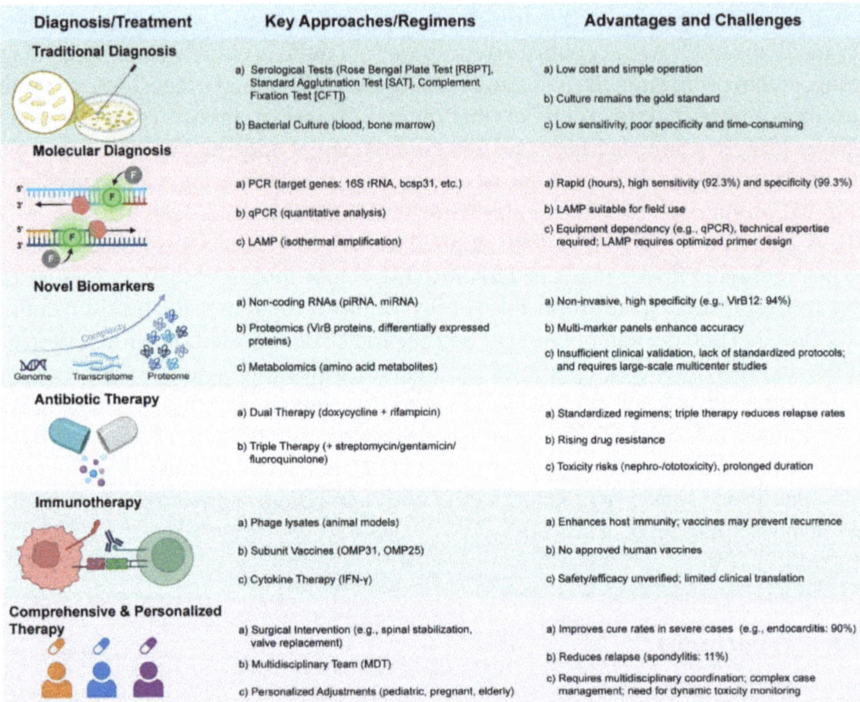


**Fig. 4.5** Diagnosis and therapy of Brucellosis. Reproduced with permission from Ref. [63]

and must be employed cautiously. Co-trimoxazole combination with rifampicin or gentamicin is recommended in children, and quinolones must be utilized cautiously as monotherapy. Brucellosis treatment is usually avoided in production animals, and affected animals are typically culled. Close monitoring along with follow-up are essential for successful treatment as well as prevention of relapse. However, relapse rates and treatment failure in mild cases of brucellosis are 5–15% common, highlighting the necessity for continuous monitoring along with frequent serological testing for a year. Emergence of the multidrug-resistant strains of *Brucella* worldwide in areas where disease is endemic has been associated with inappropriate antimicrobial usage. Antibiotics utilization in livestock contributes to the current issue, causing public health concern and also limiting available treatments. Antimicrobial susceptibility routine testing is crucial for effective management of brucellosis [61]. Currently, for brucellosis no human vaccine exists and brucellosis occurrence in the region is directly related to animal brucellosis status. Therefore, it is necessary to implement interventions that can control infection in animal reservoirs and prevent transmission of disease from animals to humans, offering more effective as well as economically viable approaches in management of disease rather than focusing on human population alone [62].

## 4.12 Prevention and Control

WHO proposed general strategies and Mediterranean Zoonoses Control Program in 1998 for eradicating brucellosis in animals. The program and strategies are as follows.

(1) preventing disease prolongation in animals along with monitoring regions and herds which are free from brucellosis.
(2) utilizing diagnostic tests for infected animals' identification and their removal via slaughter programs for making zones and herds brucellosis-free.
(3) implementing massive vaccination programs for decreasing disease prevalence. However, dairy product pasteurization is considered a significant safety process in endemic areas. Consumption of raw milk along with dairy products and undercooked or raw animal products including bone marrow must be avoided. Occupational contact to *Brucella* might be prevented through good hygiene as well as using protective equipment or clothing. Employment of safety measures is crucial for preventing inhalation, skin contamination, or accidental ingestion of pathogens while facilitating at birth, performing necropsy, and animal butchering. Furthermore, in management of aborted fetus, its membranes as well as fluids need special precaution.

Currently, there is no protective and safe human vaccine for preventing brucellosis; therefore, vaccination of animals is essential factor for controlling and eradicating human and animal brucellosis. Ideal vaccine targeting *Brucella* must:

(1) prevent brucellosis in both genders,
(2) not induce illness in animals that are immunized,
(3) control abortion,
(4) provide prolonged protection with single dose,
(5) not disrupt serological tests based on LPS,
(6) biologically stable without presenting virulence reversion threat,
(7) harmless for humans,
(8) not result in contamination of byproducts from vaccinated animals.

### *4.12.1 Live, Attenuated Vaccines*

Currently, animal vaccination targeting *Brucella* infection is typically performed by administering live-attenuated smooth *Brucella* strains such as *B. abortus* S19 strain, *B. melitensis* Rev.1 strain and *B. abortus* RB51 strain [64]. Most widely employed vaccine for preventing cattle brucellosis is *B. abortus* S19 vaccine. Owing to induction of prolonged cross-reacting immune response, S19 utilization discontinued during latter stages of brucellosis elimination program and replaced with serologically non-reactive vaccine in late1990s [65]. Despite live, attenuated vaccines provide long-lasting protection, however they have significant drawbacks:

(1) instigating abortion in animals that are pregnant,
(2) secreting in vaccinated animals' milk,
(3) pathogenic to humans,
(4) interferes with diagnostic tests based on LPS,
(5) being resistant to rifampicin that is the first antibiotic recommended for treatment of human brucellosis.

### *4.12.2 Subunit Vaccines*

Subunit vaccines emerged as favorable vaccine candidates targeting *Brucella* infection owing to safety profile in recent decades. Main benefits of subunit vaccines than live vaccines are they eliminate safety concerns, less biohazardous, avirulent, noninfectious, well defined, and nonviable. Subunit vaccines, however, cannot replicate live vaccines' immunogenicity and therefore, they are not more efficient over live-attenuated vaccines. For development of novel and effective *Brucella* vaccine, selecting immunogen with capacity to trigger suitable immune reaction (biased toward Th1) and provide high protection level is essential. Employing immunoinformatics algorithms and mapping potential B and T-cell epitopes are favorable methods for designing novel vaccine candidates. Moreover, variety of immunogens investigated by deploying bioinformatics tools for developing novel vaccine targets designed via epitope mapping [64].

## References


1. Addis M (2015) Public health and economic importance of brucellosis: a review. Public Health 5(7):68–84
2. Schelling E et al (2003) Brucellosis and Q-fever seroprevalences of nomadic pastoralists and their livestock in Chad. Prev Vet Med 61(4):279–293
3. Hosein H et al (2018) Evaluation of the general organization of veterinary services control program of animal brucellosis in Egypt: an outbreak investigation of brucellosis in buffalo. Veterinary World 11(6):748
4. Saxena N, Singh BB, Saxena HM (2018) Brucellosis in sheep and goats and its serodiagnosis and epidemiology. Int J Curr Microbiol App Sci 7(1):1848–1877
5. Godfroid J (2002) Brucellosis in wildlife. Revue Scientifique et Technique-Office international des épizooties 21(1):277–286
6. Khan MZ, Zahoor M (2018) An overview of brucellosis in cattle and humans, and its serological and molecular diagnosis in control strategies. Trop Med Infect Dis 3(2):65
7. Olsen S, Tatum F (2010) Bovine brucellosis. VetY Clin Food Anim Pract 26(1):15–27
8. Pal M et al (2017) Public health and economic importance of bovine Brucellosis: an overview. Am J Epidemiol 5(2):27–34
9. Garin-Bastuji B et al (2006) The diagnosis of brucellosis in sheep and goats, old and new tools. Small Rumin Res 62(1–2):63–70
10. Blasco JM, Molina-Flores B (2011) Control and eradication of Brucella melitensis infection in sheep and goats. Vet Clin Food Anim Pract 27(1):95–104

11. H Abd-El Halim M, AE Mohamed A, A Shalaby N (2017) Prevalence of brucellosis in buffaloes and its control measures. J Vet Med Res 24(1):84–93
12. Olsen S, Palmer M (2014) Advancement of knowledge of Brucella over the past 50 years. Vet Pathol 51(6):1076–1089
13. Seleem MN, Boyle SM, Sriranganathan N (2010) Brucellosis: a re-emerging zoonosis. Vet Microbiol 140(3–4):392–398
14. Khurana SK et al (2021) Bovine brucellosis–a comprehensive review. Veterinary Q 41(1):61–88
15. Galinska EM, Zagórski J (2013) Brucellosis in humans-etiology, diagnostics, clinical forms. Ann Agric Environ Med 20(2)
16. Mathew C et al (2015) First isolation, identification, phenotypic and genotypic characterization of Brucella abortus biovar 3 from dairy cattle in Tanzania. BMC Vet Res 11(1):156
17. Whatmore AM et al (2014) Brucella papionis sp. nov., isolated from baboons (Papio spp.). Int J Syst Evol Microbiol 64(Pt_12):4120–4128
18. Kitt E et al (2017) A case report of pediatric brucellosis in an Algerian immigrant. In: Open forum infectious diseases. Oxford University Press, US
19. Tulu D (2022) Bovine brucellosis: epidemiology, public health implications, and status of brucellosis in Ethiopia. Vet Med Res Rep 21–30
20. Moriyón I, López-Goñi I (1998) Structure and properties of the outer membranes of Brucella abortus and Brucella melitensis
21. Adem A, Duguma A (2020) Characteristics and intracellular life of Brucella organism: a review. J Microb Biochem Technol 12(3):431
22. He Y (2012) Analyses of Brucella pathogenesis, host immunity, and vaccine targets using systems biology and bioinformatics. Front Cell Infect Microbiol 2:2
23. Pal M et al (2020) Human and animal brucellosis: a comprehensive review of biology, pathogenesis, epidemiology, risk factors, clinical signs, laboratory diagnosis. Am J Infect Dis 8(4):118–126
24. Blasco JM et al (2023) A review of three decades of use of the cattle brucellosis rough vaccine Brucella abortus RB51: myths and facts. BMC Vet Res 19(1):211
25. Shi J-F et al (2021) Seroprevalence of brucellosis in buffalo worldwide and associated risk factors: a systematic review and meta-analysis. Front Vet Sci 8:649252
26. Dorneles EM, Sriranganathan N, Lage AP (2015) Recent advances in Brucella abortus vaccines. Vet Res 46(1):76
27. Castaño-Zubieta MR et al (2021) Evaluation of the safety profile of the vaccine candidate Brucella melitensis 16MΔvjbR strain in goats. Vaccine 39(3):617–625
28. Rebollada-Merino A et al (2023) Histopathologic and immunohistochemical findings in the placentas and fetuses of domestic swine naturally infected with Brucella suis biovar 2. J Vet Diagn Invest 35(3):258–265
29. van Dijk MA et al (2021) Transboundary spread of Brucella canis through import of infected dogs, the Netherlands, November 2016–December 2018. Emerg Infect Dis 27(7):1783
30. Galluzzo P et al (2021) Diagnostic findings in a confirmed outbreak of Brucella ovis infection in a traditional sheep farm in Sicily (South-Italy). Pathogens 10(11):1472
31. Socorro Ruiz-Palma Md et al (2021) Comparative proteomic analysis of outer membrane vesicles from Brucella suis, Brucella ovis, Brucella canis and Brucella neotomae. Arch Microbiol 203(4):1611–1626
32. Cavaco S et al (2025) Brucella ceti in Common Dolphins (Delphinus delphis) in Portugal—characterization of first isolates. Animals 15(3):374
33. Xavier MN et al (2010) Pathogenesis of Brucella spp. Open Vet Sci J 4(1):109–118
34. Rudolf I et al (2025) Brucella microti and rodent-borne brucellosis: a neglected public health threat. Zoonoses Public Health 72(1):1–8
35. Scholz HC et al (2010) Brucella inopinata sp. nov., isolated from a breast implant infection. Int J Syst Evol Microbiol 60(4):801–808
36. Zhang J et al (2014) Prediction and control of brucellosis transmission of dairy cattle in Zhejiang Province, China. PLoS ONE 9(11):e108592

37. Wang Y et al (2024) A scoping review on the epidemiology and public significance of Brucella abortus in Chinese dairy cattle and humans. One Health 18:100683
38. Sherasiya R (2024) Global prevalence of human brucellosis: a systematic review and meta-analysis
39. Ali S et al (2024) Molecular epidemiology of brucellosis in Asia: insights from genotyping analyses. Vet Res Commun 48(6):3533–3550
40. Shirzadi MR et al (2021) The incidence and geographical distribution of brucellosis in Iran using geographic information system and prediction of its incidence in 2021. J Prev Med Hyg 62(3):E635
41. Rankova R, Balieva G (2024) Distribution and prevalence of brucellosis outbreaks in Europe in the period 2002–2022. Sci Work Ser C Vet Med 70(1)
42. Neta AVC et al (2010) Pathogenesis of bovine brucellosis. Vet J 184(2):146–155
43. Xiao Y et al (2022) Inflammatory mechanism of Brucella infection in placental trophoblast cells. Int J Mol Sci 23(21):13417
44. Poester F, Samartino L, Santos R (2013) Pathogenesis and pathobiology of brucellosis in livestock. Rev sci tech 32(1):105–115
45. Megid J, Mathias LA, Robles C (2010) Clinical manifestations of brucellosis in domestic animals and humans
46. Xavier MN et al (2009) The genus Brucella and clinical manifestations of brucellosis. Ciência Rural 39:2252–2260
47. Al Dahouk, S., L.D. Sprague, and H. Neubauer (2013) New developments in the diagnostic procedures for zoonotic brucellosis in humans. Revue scientifique et technique (Int Off Epizoot) 32(1):177–188
48. Di Bonaventura G et al (2021) Microbiological laboratory diagnosis of human brucellosis: an overview. Pathogens 10(12):1623
49. Kaltungo B et al (2014) Brucellosis: a neglected zoonosis
50. Zvizdić Š et al (2006) Brucella melitensis review of the human infection case. Bosn J Basic Med Sci 6(1):15
51. Naseer A et al (2023) Brucella melitensis vaccines: a systematic review. Agriculture 13(11):2137
52. Castro JCV et al (2023) Aortic valve endocarditis caused by Brucella melitensis
53. Bosilkovski M, Keramat F, Arapović J (2021) The current therapeutical strategies in human brucellosis. Infection 49(5):823–832
54. Andrade RS et al (2024) Accuracy of serological tests for bovine brucellosis: a systematic review and meta-analysis. Prev Vet Med 222:106079
55. Souza P et al (2024) Analytical performance evaluation of a rapid immunochromatographic test for the diagnosis of bovine brucellosis based on a recombinant BP26 protein. Arquivo Brasileiro de Medicina Veterinária e Zootecnia 76(5):e13158
56. Mabe L et al (2022) Accuracy of molecular diagnostic methods for the detection of bovine brucellosis: a systematic review and meta-analysis. Veterinary World 15(9):2151
57. Yao X et al (2024) Study on the value of second-generation sequencing technology in the clinical diagnosis of osteoarticular brucellosis. J Orthop Res® 42(10):2327–2335
58. Albadrani BA, Abdel-Raheem M, Al-Farwachi MI (2024) Artificial intelligence in veterinary care: a review of applications for animal health. Egypt J Veterinary Sci 55(6):1725–1736
59. Xiao S et al (2025) Review of applications of deep learning in veterinary diagnostics and animal health. Front Veterinary Sci 12:1511522
60. Franca T et al (2025) Improving bovine brucellosis diagnostics: rapid, accurate detection via blood serum infrared spectroscopy and machine learning. ACS omega
61. Qureshi KA et al (2023) Brucellosis: epidemiology, pathogenesis, diagnosis and treatment–a comprehensive review. Ann Med 55(2):2295398
62. Tumwiine J, Robert G (2017) A mathematical model for treatment of bovine brucellosis in cattle population. J Math Model 5(2):137–152
63. Lu P et al (2025) Progress in brucellosis immune regulation inflammatory mechanisms and diagnostic advances. Eur J Med Res 30(1):830

64. Golshani M, Buozari S (2017) A review of brucellosis in Iran: epidemiology, risk factors, diagnosis, control, and prevention. Iran Biomed J 21(6):349
65. Tabar GH, Jafari A (2014) Preventive and control programs for brucellosis in human and animals. J Zoonoses 1(1)

# Chapter 5
# Chitosan Nanoparticles for Brucellosis Treatment

**Abstract** Brucellosis presents a notable challenge due to the intracellular persistence and adaptability of Brucella species within macrophages, complicating effective treatment strategies. Nanotechnology-based approaches predominantly CSNPs offer promising strategy for targeted drug delivery and improved therapeutic efficiency. CS and its derivatives serve as advantageous vaccine adjuvants and delivery vehicles, attributed to their significant positive charge and capacity to traverse biological barriers while binding to both drugs and cells. CSNPs have applications in mucosal delivery, especially in nasal administration, serving as mucosal vaccine adjuvants in various species including cattle, rabbits, mice, pigs, and chickens. CSNPs loaded with *Brucella abortus* (*B. abortus*) malate dehydrogenase (Mdh) induced proinflammatory cytokine production and systemic Immunoglobulin A (IgA) in mice through intranasal administration, demonstrating efficacy in eliciting strong immune responses, particularly against *B. abortus* infections. The evaluation of antimicrobial properties of CSNPs containing gentamicin through minimum inhibitory concentration (MIC) and minimum bactericidal concentration (MBC) assays demonstrated the efficacy of this formulation in eradicating both *B. abortus* and *Brucella melitensis* (*B. melitensis*) strains. The intraperitoneal administration of N-trimethyl chitosan/urease NPs enhances specific immune responses and increases protective efficacy against *B. abortus* and *B. melitensis*. The findings highlight the potential of CSNPs as a novel platform for brucellosis treatment, offering advantages compared to conventional treatment methods.

**Keywords** *Brucella melitensis* · Drug delivery · Immunoglobulin · MBC · MIC

## 5.1 Introduction

Traditionally, antibiotics have been employed as first-line treatment for infectious bacterial diseases. Nevertheless, emergence of antibiotic resistance in a wide range of bacteria has made conventional treatments less effective in both animals and humans. Hence, pervasive antibiotic resistance also instigates challenges and

 
A. Haider et al., *Chitosan Nanoparticles for Brucellosis Treatment*, Springer Series in Biomaterials Science and Engineering 21, https://doi.org/10.1007/978-3-032-16292-2_5

economic burdens in addition to making conventional treatment methods ineffective. The challenge of addressing bacterial infections due to antibiotic resistance necessitates the investigation of alternative strategies, including the development of novel antimicrobial agents, non-antibiotic treatment modalities, and combination therapies. Addressing challenges posed by antibiotic resistance is crucial for mitigating complications and reducing economic losses associated with bacterial infectious diseases. The unique physicochemical properties of nano-carriers, characterized by a high surface area-to-volume ratio, facilitate extended antibiotic release and enhance antibacterial efficacy in intracellular environments. A variety of organic, inorganic, and semi-organic nanomaterials including solid lipid NPs (SLNPs), polymersomes, liposomes, nanostructured lipid carriers (NLCs), quantum dots, carbon-based nanomaterials, silica NPs, and metal/metal oxide NPs are investigated as nano-carriers for antibiotics [1].

CSNPs exhibit exceptional physicochemical, biological, and antimicrobial capabilities, categorizing them as natural materials. This characteristic renders them environmentally sustainable and ensures their bioactivity is safe for human use. Due to these unique properties, CSNPs are utilized in various applications. CSNPs, due to their mucoadhesive and biological attributes, can enhance transmucosal permeability, thereby increasing transport through the paracellular pathway of NPs and initiating structural reorganization of proteins connected to tight junctions. Semiconductor nanocrystals, commonly referred to as quantum dots, serve as highly effective fluorescent probes in a range of biomedical applications. Despite the success in utilizing this nanocrystal, concerns persist regarding cytotoxicity associated with its heavy metal composition. The non-toxic characteristics of CSNPs render them significant in this context. The anticancer activity of CSNPs may be attributed to their small size. Reduced particle size improves the specific surface-area and surface-volume ratio, subsequently improving dissolution and bioavailability of CS.

The potential employment of CSNPs as carriers has facilitated development of a broad range of colloidal delivery vehicles. CSNPs possess the capability to traverse biological barriers, thereby safeguarding macromolecules from degradation within biological environments. It may also facilitate the delivery of macromolecules or pharmaceuticals through a regulated release mechanism to the designated site. CS has the capacity to create a gel when it interacts with anions, leading to the formation of beads. This property facilitates the employment of CS in the realm of drug delivery. Nonetheless, the considerable dimensions of the beads (1–2 mm) continue to limit their applicability. The diminutive dimensions of CSNPs augment their efficacy in interfacial interactions with cell membranes, thereby promoting the cellular uptake of small particles through the process of endocytosis. A multitude of studies has indicated the potential of CSNPs to improve the bioavailability of drugs, thereby modifying their pharmacokinetics and safeguarding encapsulated substances [2]. A study demonstrated that carbon dots (CDs)/CS-doped lanthanum oxide ($La_2O_3$) nanorods exhibited inhibition zones measuring 3.55 mm and 4.15 mm at low and high concentrations, respectively, against the gram-negative bacteria *E. coli*. The enhancement of bactericidal efficacy through CS doping can be attributed to the critical role of electrostatic interactions between cationic molecules and the negatively charged walls of bacterial cells [3].

In addition to being utilized as an oral delivery carrier, CSNPs can also be used in mucous membrane systems as nasal and pulmonary routes for delivery of proteins and peptides. The quest for natural antimicrobials to reduce the reliance on synthetic chemicals has resulted in the development of CS and CSNPs. The interaction between positively charged CS and the negatively charged phospholipids of the plasma membrane leads to alterations in cell permeability, ultimately resulting in the leakage of cellular components and subsequent cell death. The capacity of CS to chelate metal ions may be a contributing factor to its antimicrobial properties. Reports indicate that CS is capable of penetrating the cell wall and binding to DNA, thereby inhibiting mRNA synthesis [2].

## 5.2 Nano Drug Delivery in Treatment of *Brucella*

Vaccination serves as a crucial mechanism for controlling and preventing the spread of infectious diseases by eliciting a robust protective immune response. This process markedly enhances the lifespan and diminishes mortality rates. Traditional vaccines, such as live attenuated vaccines, can elicit robust immune responses; however, they also raise significant safety concerns. Although inactivated vaccines are considerably safer, their immune responses tend to be suboptimal. Subunit vaccines, which may consist of one or more protective antigens, exhibit a tendency toward low immunogenicity. Nucleic acid vaccines encompass both mRNA vaccines and DNA vaccines. DNA vaccines are relatively easy to construct and can be developed on a large scale; however, they exhibit low immunogenicity due to a low transfection rate and subsequent protein expression. mRNA vaccines exhibit transient expression and are directly translated within the cytoplasm, eliminating the need for insertional mutagenesis. They demonstrate high transfection efficacy, rapid production, relatively low cost, and remarkable versatility in vaccine design. However, they exhibit instability, are susceptible to bodily degradation, and lack an effective means of delivery. A multitude of studies has substantiated that CS and its derivatives serve as advantageous vaccine adjuvants and delivery vehicles, attributed to their significant positive charge and their capacity to traverse biological barriers while binding to both drugs and cells (Fig. 5.1) [4].

CSNPs have been employed in mucosal delivery, particularly in nasal delivery such as mucosal vaccine adjuvants in rabbits, mice, pigs, chickens, and cattle. Intranasal immunization with CSNPs induced both humoral and cellular immune responses in many studies [6]. CSNPs loaded with *B. abortus* malate dehydrogenase (Mdh) instigated pro-inflammatory cytokine production as well as systemic Immunoglobulin A (IgA) in mice via intranasal administration, indicating effectiveness for robust immune responses, mainly in tackling *B. abortus* infections. These NPs activated immune cells as well as pathways associated with mucosal immunity

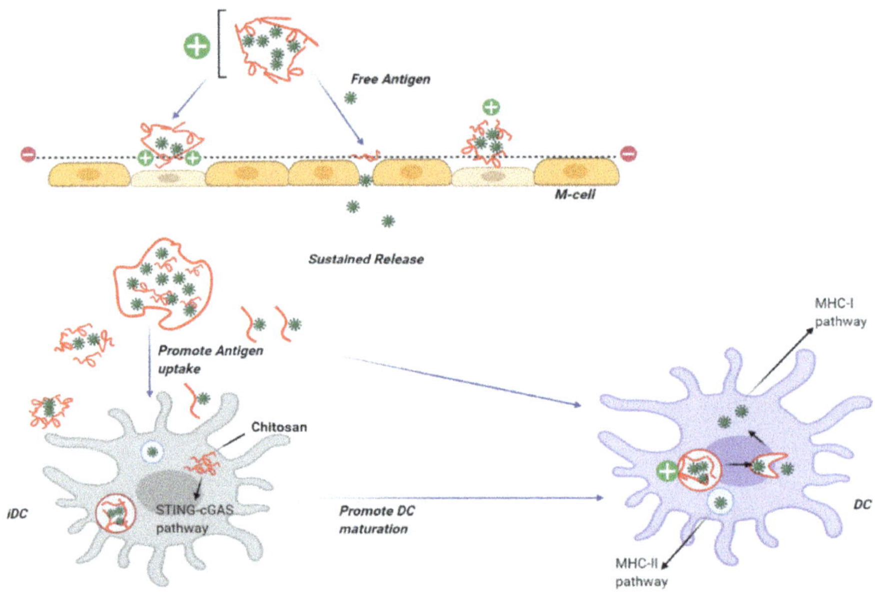


**Fig. 5.1** Use of chitosan as an adjuvant for vaccine delivery. CH can be dissolved in acid to provide a positive charge on the particle surface. CH NPs containing antigens can adhere to the cell surface in the nasal mucosa, giving a prolonged residence time. CH can open the tight junctions and allow the transfer of free antigens through the paracellular route, as well as the transcytosis of the packaged antigens via M-cells. CH becomes insoluble at physiological pH, providing slow release of encapsulated antigens and uptake by dendritic cells DCs. Following uptake by DCs, CH activates the STING-cGAS pathway, resulting in DC maturation. After the insoluble particles have been taken up by cells via endocytosis, CH becomes soluble at the acidic pH, and the antigens escape from the lysosomes into the cytoplasm for cross presentation through the MHC-I pathway. Antigens taken up by DCs can be routed to the MHC-II pathway. Reproduced with permission from Ref. [5]

in nasal-associated lymphoid tissue (NALT), resulting in significant formation of specific IgA (Fig. 5.2) [7].

The intranasally administered *B. abortus* antigen and CSNPs in mice elicited a mixed Th1/Th2 response at four weeks post-infection (wpi) and subsequently resulted in a Th2 response at six wpi. The utilization of these delivery systems resulted in the induction of antigen-specific IgA, driven by a Th2-polarized immune response, through the administration of 3 *B. abortus* antigens (Mdh, Omp19, and Omp10) encapsulated within CSNPs. The amalgamation of these robust immunogenic antigens resulted in the production of IgG that is specific to each antigen type, alongside IgA that is specific to Mdh (Fig. 5.3) [6]

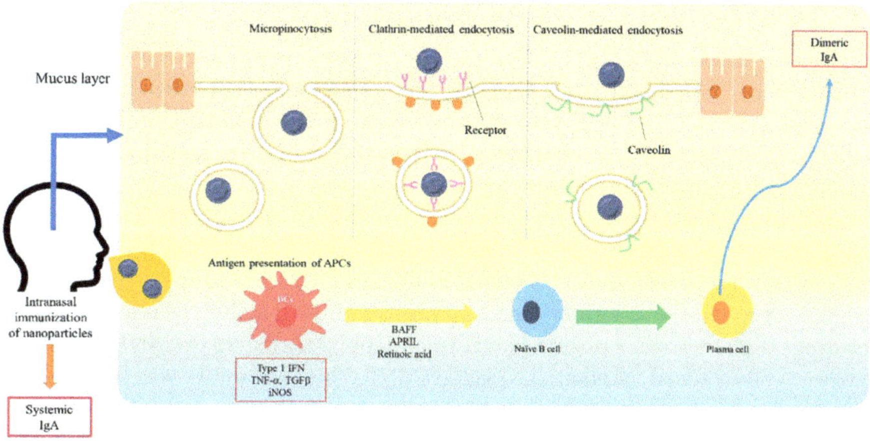


**Fig. 5.2** Schematic diagram of interaction between nanoparticles and a mucosal membrane, and the induction of mucosal immunity. At the mucosal membrane, the transport of nanoparticles is considered as mediated by caveolin-1, clathrin, and micropinocytosis into M cells. After antigen presentation, B cells migrate from the NALT to the cervical lymph nodes. IgA-producing plasma cells subsequently produce dimeric or polymeric forms of IgA. Dimeric IgA becomes secretory IgA on the epithelial cells of the mucosal membranes and is released into the mucosal tract. Reproduced with permission from Ref. [6]

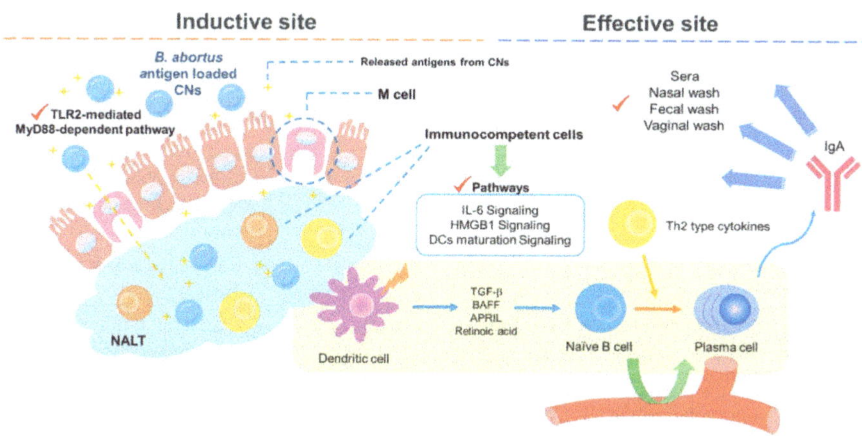


**Fig. 5.3** Schematic diagram of the suggested pathways of the immune response induced by *B. abortus* recombinant protein loaded CNs (chitosan nanoparticles). Loading *B. abortus* antigen, Mdh, into the CNs induced enhanced transport of Mdh in the in vitro M cell model and CNs–Mdh triggered activation of HMGB1, IL-6, and DCs maturation signaling within NALT in BALB/c mice and elicited Th2-related response with production of IgA at 6wpi (weeks post infection). Reproduced with permission from Ref. [6]

## 5.3 Drug Loaded CSNPs for *Brucella* Infection

CSNPs were employed for delivery of gentamicin into Brucella-infected macrophages to inspect applicability of CSNPs in intracellular drug delivery as carriers. Assessment of antimicrobial properties of CSNPs containing gentamicin utilizing minimum inhibitory concentration (MIC) and minimum bactericidal concentration (MBC) assays revealed effectiveness of this formulation in elimination of both *B. abortus* and *B. melitensis* strains. Nano-formulation indicated higher antibacterial activity as compared to free drug which was possibly because of sustained drug release in former state, whereas duration of activity was low in case of free drug. Higher efficacy of entrapped drug might be attributed to CSNPs' ability to penetrate cells. Indeed, gentamicin-loaded CSNPs' bioavailability was found higher in comparison to free gentamicin. In the presence of free drug, cells are exposed to high drug concentrations for a brief duration. Conversely, with gentamicin-loaded CSNPs, the drug remains available for an extended period, thereby more effectively reducing bacterial load [8]. Table 5.1 mentions a few of several studies on drug delivery modes of nanoparticles with antibiotics [9].

Mucosal immunizations (mainly oral immunization) are topic of great attention owing to convenience and simplicity. The advantages of oral immunization encompass the management of infections at the pathogen's entry point, the stimulation of both systemic and mucosal immune responses, and the elimination of the need

**Table 5.1** Studies illustrating the drug delivery modes with antibiotics according to [10–14]

| Ref. | Type of delivery system | Biomaterial/ polymer | Drug inoculated | Conclusions |
|---|---|---|---|---|
| Goodson et al. [10] | Hydrogel | Chitosan | 5% Tetracycline | • Great antibacterial activity against Gram-negative and Gram-positive bacteria<br>• Can be used as wound dressing |
| Greenstein and Polson [11] | Nanospheres | Chitosan | Ciprofloxacin | • Excellent antibacterial activity against *E. coli* |
| Agnihotri et al. [12] | Polyelectrolyte films | Chitosan + Alginate | Clindamycin | • Good drug delivery system for periodontal therapy |
| Ghafar et al. [13] | Nanoparticle | Chitosan + Dextran | IL-17RB siRNA and doxorubicin (DOX) | • Co-delivery of IL17RB siRNA and DOX have shown excellent results in the treatment of breast cancer |
| Khattak et al. [14] | Intracanal medicament | Chitosan | Calcium hydroxide + Triple antibiotic paste | • Highly effective against *E. faecalis* and *C. albicans* to treat endodontic infections |

Reproduced with permission from Ref. [9]

for needles. However, the process of oral immunization presents significant challenges due to its remarkably low bioavailability. The formulation of oral vaccines necessitates the navigation of various challenges, including limited permeability of large molecules, lack of drug lipophilicity, and rapid enzymatic inactivation or degradation within the gastrointestinal tract. In addressing these challenges, a range of polymeric NPs have been examined as delivery systems for the intestine, potentially safeguarding the cargo from adverse conditions that could compromise vaccine bioactivity. Particles made from bioadhesive substances such as N-trimethyl chitosan (TMC), which consistently exhibit stable positive charges despite pH fluctuations, can extend the residence time of vaccines in the intestine and enhance their immunogenicity and permeation. Furthermore, TMC NPs have the capacity to promote maturation of dendritic cells (DCs), thereby demonstrating inherent adjuvant characteristics. Consequently, TMC NPs serve as an advantageous delivery system for immunization due to these characteristics.

Polymeric NPs encapsulating antigen were employed for oral vaccination. For systemic vaccination, CS and TMC NPs are effective vaccine delivery systems. Oral immunization of Omp31-loaded TMC NPs demonstrated comparable protection against *B. melitensis* 16M challenge in mice when compared with Rev.1 vaccine strain, evaluated 4 weeks after challenge. Oral administration of TMC/Omp31 NPs triggers significant Th1–Th17 cellular-mixed immune response. Oral immunization provided higher protection than intraperitoneal immunization. Omp31 vaccination induced robust production of antigen-specific cytokine-secreting cells in spleen, which can be enhanced further after TMC/Omp31 NPs oral vaccination [15].

The identification of a viable vaccine candidate is essential for the development of effective and safe subunit vaccines aimed at combating brucellosis. FliC protein was incorporated into mannosylated chitosan nanoparticles (MCN) developed as a method for targeted vaccine delivery. The protective efficacy of the FliC protein was noted in BALB/c mice. The administration of FliC and FliC-MCN conferred notable protection in murine models against infections caused by *B. abortus* 544 and *B. melitensis* 16M, in contrast to the MCN groups and phosphate-buffered saline (PBS). The findings indicate that the FliC protein may serve as a promising antigen candidate for the development of a subunit vaccine targeting *B. abortus* and *B. melitensis*. Moreover, MCN can serve both as an adjuvant and as a targeted system for vaccine delivery [16].

## 5.4 In Vivo Evaluation and Mechanistic Insights of CSNPs

CS and calcium phosphate (CaP) have been explored as nano-carriers and have displayed promising biocompatibility and safety profiles. CaP/CS NPs are emerging as favorable candidates in vaccine development owing to exceptional properties as biocompatible as well as biodegradable materials, which make them appropriate for medical applications. Combination of CS and CaP enables vaccine antigens-controlled release to protect them from deterioration and ensure effective targeting.

Sonicated *B. melitensis* whole-cell antigens loading in CaP/CS NPs are safe and favorable as active immunogens for inducing protection against *B. melitensis* [17].

Guinea pigs were experimentally infected utilizing virulent field *B. melitensis* biovars 3, that is a common type of *Brucella* isolated from animals as well as humans, were used to assess effectiveness of in situ pH-responsive curcumin-loaded niosome hydrogel and doxycycline-loaded chitosan–sodium alginate NPs. The bacteriological examination of the spleens from Guinea pigs infected with Brucella, conducted two weeks post-infection, revealed smooth cultures characterized by an abundance of colonies. These colonies exhibited a transparent, convex morphology with a brilliant surface and well-defined borders, transitioning to a honey hue upon exposure to transmitted light. Every culture was classified as *B. melitensis* biovar 3. All control-negative Guinea pigs exhibited negative results in bacteriological assessments. The histological structure of the spleen in tissue sections from control animals, which were not subjected to infection, exhibited a normal appearance. The histopathological examination of the spleen in the infected groups revealed the presence of early granulomatous inflammatory responses, predominantly characterized by epithelioid macrophages, alongside a minor presence of neutrophils within the spleen's architecture. Diffuse congestion was markedly observed in the red pulp of the spleen. Neither necrosis nor giant cells were observed. The in vivo experimentation confirmed that the combination of doxycycline-loaded chitosan–sodium alginate NPs and in situ pH-responsive curcumin-loaded niosome hydrogel represents an exceptional formulation for the treatment of brucellosis and the reduction of splenic count [18].

The recombinant protein *Brucella* urease has been identified as a factor that elicits protective responses against *B. melitensis* and *B. abortus* in BALB/c mice. TMC is favored for a range of biomedical applications as a regulated drug delivery mechanism for vaccines, biomolecules, and therapeutics. The simultaneous administration of TMC alongside the antigen enhances the immune response and provides greater protection compared to the administration of the antigen in isolation. Highly pathogenic strains of *B. abortus* and *B. melitensis* were utilized to test the resilience of vaccinated animals. The production of IFN-γ and IL-12 was elevated in mice that received intraperitoneal (i.p.) and oral vaccinations with TMC/Urease NPs. The intraperitoneal administration of TMC/Urease NPs demonstrates a superior immune response when contrasted with oral administration. The elevated Th1 and Th2 immune responses observed with the intraperitoneal administration of TMC/Urease NPs may account for the notable protection against challenges posed by *B. abortus* and *B. melitensis*, in contrast to the level of protection achieved through oral administration of TMC/Urease NPs. Consequently, the administration of TMC/Urease NPs via the intraperitoneal route augments specific immune responses and improves protective efficacy [19].

## References


1. Alavi M, Nokhodchi A (2023) Micro-and nanoformulations of antibiotics against Brucella. Drug Discov Today 28(12):103809
2. Divya K, Jisha M (2018) Chitosan nanoparticles preparation and applications. Environ Chem Lett 16(1):101–112
3. Rasool S et al (2023) Efficient dye degradation and antibacterial activity of carbon dots/chitosan-doped $La_2O_3$ nanorods: in silico molecular docking analysis. ACS Omega 8(28):25401–25409
4. Gong X et al (2022) Chitosan-based nanomaterial as immune adjuvant and delivery carrier for vaccines. Vaccines 10(11):1906
5. Boroumand H et al (2021) Chitosan-based nanoparticles against viral infections. Front Cell Infect Microbiol 11:643953
6. Shim S, Yoo HS (2020) The application of mucoadhesive chitosan nanoparticles in nasal drug delivery. Mar Drugs 18(12):605
7. Omidian H et al (2024) Chitosan nanoparticles for intranasal drug delivery. Pharmaceutics 16(6):746
8. Razei A et al (2019) Gentamicin-loaded chitosan nanoparticles improve its therapeutic effects on brucella-infected J774A. 1 murine cells. Galen Med J 8:e1296
9. Mascarenhas R, Hegde S, Manaktala N (2024) Chitosan nanoparticle applications in dentistry: a sustainable biopolymer. Front Chem 12:1362482
10. Goodson J et al (1985) Periodontal disease treatment by local drug delivery. J Periodontol 56(5):265–272
11. Greenstein G, Polson A (1998) The role of local drug delivery in the management of periodontal diseases: a comprehensive review. J Periodontol 69(5):507–520
12. Agnihotri SA, Mallikarjuna NN, Aminabhavi TM (2004) Recent advances on chitosan-based micro-and nanoparticles in drug delivery. J Control Release 100(1):5–28
13. Ghafar H et al (2020) Development and characterization of bioadhesive film embedded with lignocaine and calcium fluoride nanoparticles. AAPS PharmSciTech 21(2):60
14. Khattak S et al (2019) Applications of cellulose and chitin/chitosan derivatives and composites as antibacterial materials: current state and perspectives. Appl Microbiol Biotechnol 103(5):1989–2006
15. Abkar M et al (2017) Oral immunization of mice with Omp31-loaded N-trimethyl chitosan nanoparticles induces high protection against *Brucella melitensis* infection. Int J Nanomed, 8769–8778
16. Sadeghi Z et al (2020) Mannosylated chitosan nanoparticles loaded with FliC antigen as a novel vaccine candidate against *Brucella melitensis* and *Brucella abortus* infection. J Biotechnol 310:89–96
17. Mohammed HS, Al-Rasheed AA, Noomi BS (2025) Preparation and characterization of caphosphate/chitosan nanoparticles containing antigen of brucella melitensis: safety assessment in vivo (in rat model). Assiut Vet Med J 71(185):398–412
18. Abo El-Ela FI et al (2020) Treatment of brucellosis in guinea pigs via a combination of engineered novel pH-responsive curcumin niosome hydrogel and doxycycline-loaded chitosan–sodium alginate nanoparticles: an in vitro and in vivo study. AAPS PharmSciTech 21(8):326
19. Abkar M et al (2018) Intraperitoneal immunization with Urease loaded N-trimethyl Chitosan nanoparticles elicits high protection against *Brucella melitensis* and *Brucella abortus* infections. Immunol Lett 199:53–60

# Chapter 6
# Conclusion, Challenges, and Future Prospective

**Abstract** This chapter integrates findings on CSNPs as a novel nanoplatform for addressing brucellosis, highlighting essential conclusions, ongoing challenges, and prospective avenues for future research. CSNPs, characterized by their biocompatibility, cohesiveness, antimicrobial properties, and encapsulation capacity, represent a promising approach for targeted and sustained delivery against intracellular Brucella spp. Various challenges persist, including stability, accurate targeting of macrophages, enhancement of intracellular delivery to phagolysosomal compartments, and necessity for synthesis process that is scalable and reproducible in accordance with regulatory standards. The upcoming environment encompasses the engineering of functionalized CSNPs employing ligand-mediated targeting, the development of stimuli-responsive release systems, the formulation of immunomodulatory agents, and the establishment of hybrid nanocomposites, all aimed at improving therapeutic efficiency and addressing the challenge of antibiotic resistance. Advancements in in vivo imaging, nanotoxicology, and clinical nanomedicine will significantly improve the development of next-generation chitosan-based nanotherapeutics for the efficient management of brucellosis.

**Keyword** CSNP · Functionalized · In-vivo · Nanotoxicology · Stimuli responsive

A zoonotic disease caused by Brucella species has continually constituted serious issue to livestock as well as public health globally, imposing serious implications on agricultural economies and human health in different regions. The persistence of disease can be reasonably linked to intracellular lifestyle of Brucella bacteria, which empowers these pathogens to consistently evade immune mechanisms of host and become resident within macrophages, hence making elimination challenging. Conventional antibiotic regimens, including doxycycline with rifampicin and most often with streptomycin or gentamicin, are confronted with many disadvantages affecting their effectiveness. While for prophylaxis, there exist available live attenuated vaccines for animals, e.g., S19 and RB51, they are also associated with disadvantages including incompatibility, serological diagnostic challenges, safety challenges, and abortion risk in pregnant animals. These problems highlight the need for novel



A. Haider et al., *Chitosan Nanoparticles for Brucellosis Treatment*, Springer Series in Biomaterials Science and Engineering 21, https://doi.org/10.1007/978-3-032-16292-2_6

methods to enhance drug delivery, improve therapeutic effectiveness, and ensure the safety of interventions used to combat this infection. Nanotechnology has generated significant interest due to shortcomings of current approaches for treating and preventing infections, as well as developing new methods for managing severe brucellosis. A promising nanotechnology infrastructure developed for site-specific release of antimicrobial agents against intracellular pathogens like Brucella is offered by natural polysaccharide chitin-derived chitosan nanoparticles (CSNPs), which offer previously unprecedented collection of desirable characteristics, particularly their multi-faceted usability, biodegradability, muco-adhesion, and intrinsic antimicrobial characteristics. Entrapment of drug entities within interiors of CSNPs forestalls early degradation, hence enabling site-specific release to infected cells. Further, surface modification of CSNPs via selective ligands or antibodies, in combination with in *vitro* and in *vivo* studies, enables site-specific targeting of infected macrophages, hence permitting simultaneous augmentation of site-specific concentration of drug at site of infection and forestalling any ensuing systemic toxicity. Additionally, it is worth noting that CS itself possesses immunomodulatory properties that may fortify host defenses and exhibit intrinsic antibacterial effects by compromising the structural integrity of microbial cell membranes, thereby providing dual benefit in fight against such infections. The profoundly promising solution of CSNPs is their potential to improve treatment outcomes, diminish relapse rates, and combat rising antibiotic resistance, which is conclusively validated by synergistic combination of targeted delivery, improved pharmacokinetics, and innate bioactivity.

The complete clinical translation and broad use of CSNPs in treatment of brucellosis are hindered by various hurdles, despite encouraging preclinical outcomes. CS is widely acknowledged as safe for medical applications; however, its suitability as a nano-therapeutic necessitates more thorough examination of potential organ accumulation, dose-responsive toxicity, hemocompatibility, and immunogenicity, all of which may be affected by differences in chitosan's molecular mass and degree of deacetylation. Ensuring uniformity in NP dimensions, morphology, and drug encapsulation efficiency presents significant technical challenges, particularly when transitioning to large-scale manufacturing. CSNPs' bioavailability may be restricted by biological barriers, such as reticuloendothelial system's swift clearance, and their pharmacokinetics or immunogenicity may be inadvertently altered by surface modifications intended to enhance targeting capability. Additionally, CSNPs possess inherent limitations against antibiotic resistance mechanisms, and there is, therefore, a need to consider combinations with other treatments or co-encapsulation techniques. Considering challenges regarding regulation include a lack of standard regulation submitted to NPs as therapeutic agents, and consideration of financial burden are also major obstacles before clinical treatment. Acceptance by physicians and patients will be contingent on evidence of safety, efficacy, and cost-effectiveness from well-designed clinical studies.

The prospects of CSNPs in brucellosis therapy agent seem to be promising, with ongoing research targeting the current limitations. The ultimate goal of improvement in NPs engineering is the development of more compliant, uniform, and stable CSNPs that will release therapeutic payloads in response to stimuli at infected sites.

Such targeted delivery systems with ligands (or antibodies) could increase antibiotic accumulation at infected sites and minimize side effects. CSNPs undertake novel approaches of combination therapies by enabling the co-delivery of antibiotics with immunomodulators or other resistance modifiers to improve treatment outcomes. CSNPs can improve the development of oral or intranasal vaccine forms by enabling co-delivery as vaccine carriers, thus improving access to immunization in resource-limited settings. The integration of innovative strategies in personalized medicine with sophisticated diagnostic technologies presents opportunities for more customized therapies and predictive healthcare. Comprehensive preclinical and clinical investigations, clearly defined regulatory frameworks, and efficient manufacturing processes are essential for successful implementation in clinical applications. The improvement in the management and prevention of brucellosis through CSNPs vaccines and therapies in highly endemic regions may still be beneficial, contingent upon the effective resolution of specific challenges. The effectiveness of treatment, recurrence, and resistance, along with initiatives for global eradication, present opportunities for enhancement.

In summary, utilization of CSNPs represents a highly intricate and groundbreaking methodology that holds substantial promise for both treatment and prevention of brucellosis. Although there remains a plethora of persistent challenges that span across technical, biological, and regulatory domains, the continuous advancement in research and technology is strategically positioned to overcome these formidable obstacles, thereby paving the way for clinical application and integration of therapies that are based on CSNPs. The potential for synergistic combinations with both established and novel medical strategies is considerable, suggesting that such integrative approaches could significantly improve patient care outcomes while simultaneously advancing the global efforts aimed at combating the prevalence of brucellosis.

GPSR Compliance
The European Union's (EU) General Product Safety Regulation (GPSR) is a set of rules that requires consumer products to be safe and our obligations to ensure this.

If you have any concerns about our products, you can contact us on

ProductSafety@springernature.com

In case Publisher is established outside the EU, the EU authorized representative is:

Springer Nature Customer Service Center GmbH
Europaplatz 3
69115 Heidelberg, Germany

www.ingramcontent.com/pod-product-compliance
Ingram Content Group UK Ltd.
Pitfield, Milton Keynes, MK11 3LW, UK
UKHW020633260926
475829UK00003BA/60

* 9 7 8 3 0 3 2 1 6 2 9 1 5 *